Anne Meyer

Milk Run Design

Definitions, Concepts and Solution Approaches

WISSENSCHAFTLICHE BERICHTE

Institut für Fördertechnik und Logistiksysteme
am Karlsruher Institut für Technologie (KIT)

BAND 88

Milk Run Design

Definitions, Concepts and Solution Approaches

by
Anne Meyer

Scientific
Publishing

Dissertation, Karlsruher Institut für Technologie (KIT)
Fakultät für Maschinenbau, 2015
Referenten: Prof. Dr.-Ing. Kai Furmans, Prof. Dr. Stefan Nickel

Impressum

KIT Scientific Publishing

Karlsruher Institut für Technologie (KIT)
KIT Scientific Publishing
Straße am Forum 2
D-76131 Karlsruhe

KIT Scientific Publishing is a registered trademark of Karlsruhe Institute of Technology. Reprint using the book cover is not allowed.

www.ksp.kit.edu

Print on Demand 2017 – Gedruckt auf FSC-zertifiziertem Papier

ISSN 0171-2772
ISBN 978-3-7315-0566-2
DOI: 10.5445/KSP/1000057833

Kurzfassung

Milk Runs in Zuliefernetzwerken – in dieser Arbeit als fahrplangebundene Sammeltouren verstanden – sind das hauptsächlich genutzte Transportkonzept in der japanischen Automobilindustrie, dem Ursprung schlanker Produktionsprinzipien. Trotz der Vorteile, die Milk Runs auf regelmäßig bedienten Zulieferrelationen bieten, haben sich in Europa andere Transportkonzepte entwickelt und durchgesetzt. Hauptgrund hierfür ist die unterschiedliche Gestalt der Zuliefernetzwerke. Jedoch würde ein Netzwerk, das unterschiedliche Transportkonzepte unter Berücksichtigung von Milk Runs kombiniert, die Ausnutzung der Vorteile aller Konzepte auch in Europa erlauben.

Die Komplexität der taktischen Planung eines solchen Netzwerkes – die Auswahl des Transportkonzeptes je Zulieferer, die Zuordnung einer Frequenz je Zulieferer sowie das Design der Milk Run Touren – erfordert die Einführung einer IT-gestützten Entscheidungsunterstützung: Ziel dieser Arbeit ist es, Modelle und Verfahren zur Lösung des Milk Run Design Problems aufzuzeigen sowie Kennzahlen einzuführen, die die Leistung von etablierten Milk Run Systemen in Relation zu alternativen Transportkonzepten messen.

Dazu wird das Milk Run Design Problem ausgehend von einer Klassifikation typischer in der Praxis auftretender Transportkonzepte und einer Diskussion der Rolle von Milk Runs im Kontext schlanker Produktionssysteme definiert. Die Klassifikation der Transportkonzepte bildet die Grundlage für eine Analyse in der Literatur vorgeschlagener Ansätze hinsichtlich ihrer Anwendbarkeit auf das Milk Run Design in Zuliefernetzwerken.

Da integrierte Ansätze in der Literatur fehlen, wird in dieser Arbeit das Transport Concept Assignment and Milk Run Scheduling (TC-MRS) Modell eingeführt. Es basiert auf zyklischen Lagerhaltungsmodellen und einer pfadbasierten Formulierung des Periodischen Tourenplanungsproblems. Zusätzlich werden Konsistenzanforderungen sowie eine hinsichtlich Kapazitäten und Kosten heterogene Flotte berücksichtigt. Um kleine und mittelgroße Instanzen zur Optimalität lösen zu können, wird ein einfaches aber sehr flexibles a-priori-Spaltengenerierungsverfahren vorgeschlagen. Eine erste ausführliche Evaluation der unterschiedlichen Modellvarianten wird auf Basis kleiner, künstlich erstellter Probleminstanzen vorgenommen.

Um das Potenzial des Ansatzes im praktischen Umfeld nachzuweisen, wird es außerdem auf das Zuliefernetzwerk eines Produktionsstandortes von Bosch Dieselsysteme angewendet. Die optimale Lösung des TC-MRS Modelles zeigt ein erhebliches Kosteneinsparungspotential von bis zu 15% – zusätzlich zu den übrigen Vorteilen von regelmäßigen Milk Run Touren. Allerdings können diese Potentiale nur dann realisiert werden, wenn ein erheblicher Anteil der täglich und wöchentlich bedienten Relationen als Milk Run Kandidaten berücksichtigt werden und wenn eine heterogene Flotte zur Verfügung steht.

Als Leistungskennzahl für etablierte Milk Run Systeme wird der *p*-value Indikator vorgeschlagen: Er ist definiert als Verhältnis von Kosten für ein Milk Run Cluster zu den Kosten, die bei einer Abwicklung über alternative Transportkonzepte entstehen. Damit ist die Kosteneffizienz in Relation zu alternativen Transportkonzepten wesentlich besser abgebildet als durch die reine Messung der Fahrzeugauslastung, wie sie heute in der Praxis üblich ist. Zum Abschluss werden Auswirkungen auf die (vertraglichen) Beziehungen zwischen produzierendem Unternehmen und Logistikdienstleistern diskutiert, die eine Anwendung des TC-MRS Modelles auf das Zuliefernetzwerk nach sich ziehen.

Abstract

Supplier milk runs – understood in this thesis as regularly scheduled pick up tours serving a producing plant – are one of the main transport concepts used in the Japanese automotive industry, the origin of lean manufacturing. In Europe, in spite of the advantages of milk runs for serving regular supplier plant relations, other transport concepts have been developed and are predominantly used. The main reasons are differing characteristics of the inbound networks. However, to exploit the advantages of all concepts, a mix, including milk runs, seems appropriate for Europe.

The complexity of the tactical planning task for such a mixed network – selecting an appropriate transport concept for each supplier, assigning a frequency to each supplier and designing the milk run routes – makes an IT based decision support necessary: The goal of this thesis is to provide models and a solution method for the Milk Run Design problem and to introduce indicators assessing the performance of an established milk run system in relation to alternative transport concepts.

To this end, we define the milk run design problem starting from a classification of transport concepts typical in practice and a discussion of milk runs in the context of lean manufacturing systems. The classification of transport concepts is used as a basis for analysing approaches from literature with respect to the applicability to the milk run design problem for inbound networks.

Since an integrated approach is missing in literature, we propose the Transport Concept Assignment and Milk Run Scheduling (TC-MRS) model. It is based on cyclic inventory models and a path based formulation of the periodic vehicle routing problem. We additionally consider consistency requirements and a fleet which is heterogeneous in terms of capacity and cost. For solving small to mid sized instances to optimality, a simple but flexible a priori column generation approach is proposed. A first extensive evaluation of different model variants is conducted on small artificial instances.

To show the potential of the approach in a real world setting, we apply it to a supply network of a Bosch plant producing diesel systems. By solving the TC-MRS model to optimality for this case study, substantial cost savings of up to 15% can be reached – keeping in mind additional advantages of a regular milk run system. However, the results also show that this potential can be only realised if a considerable share of the daily and weekly supplier relations are considered as milk run candidates and if a heterogeneous fleet is available.

To measure the performance of an established milk run system, we introduce the p-value indicator: It is defined as the proportion of the cost of a milk run cluster to the cost for the execution by alternative transport concepts. This indicator better represents the cost efficiency in comparison to alternatives than the load factor of the milk run vehicles, which is commonly used as performance indicator in practice. We conclude this thesis by discussing the effects of applying the TC-MRS model on the (contractual) relationship between a producing company and its logistics service providers.

Contents

1 Introduction

"Variability in a production system
will be buffered by some combination of
1. Inventory 2. Capacity 3. Time."
(Hopp and Spearman 2001a)

Lean manufacturing concepts emerged after World War II in Japan where the car manufacturers were competing intensively on a small market and had to produce a great variety of car types in very small lots (see e.g. (Womack et al. 1990), (Monden 2012) or (Hines et al. 2004)). From a logistics point of view, this situation has driven the engineers – especially of Toyota – to organize their transports between and within their plants "at fixed times along fixed routes" (Baudin 2004). This concept – referred to as milk run concept – allows to move small quantities of a large number of different items with predictable lead times and without multiplying transport costs (see (Baudin 2004)).

Since the publication of the book "The Machine That Changed the World" of Womack et al. (1990) in the early nineties of the last century, car manufacturers all over the world adapted their production according to the lean manufacturing principles, and many of them reached a high level of implementation within their plants. Accordingly, European automotive manufacturers successfully established milk runs for in-plant transports, but – in contrast to Japanese companies – they rarely apply milk runs to their inbound traffic.

A reason why the adaptation of the milk run concept in the latter case is not straight forward is the different "gestalt" of the inbound logistics subsystems (Miemczyk and Holweg 2004) in Japan and Europe: One key characteristic of the "gestalt" is – besides the number of suppliers – the average distance between a plant and its suppliers (see (Miemczyk and Holweg 2004)). According to a study of Dyer (1996), the average distance of Toyota to its affiliated suppliers is 48 km, and 140 km to its independent suppliers. In Europe, suppliers of a car manufacturer – or a first tier supplier like our case study partner Bosch – are spread all over the continent. Hence, milk runs that cycle several times per day between clusters of suppliers and receiving plants – as common in Japan – cannot be established cost-efficiently.

The predominant transport concepts for inbound traffic in the German automotive industry are point-to-point transports and area forwarding services: In a point-to-point transport, a large shipment is brought by a freight forwarder directly without further consolidation from a supplier to a plant. Area forwarding services are a special form of groupage services. In this case, a producing company gives all transport orders from a certain area to a certain location to a logistics provider and leaves it to the subcontractor to plan and execute these transports (see (Projektgruppe Standardbelieferungsformen 2008)). Usually, in these cases, the suppliers and logistics service providers are informed only one day before the pickup needs to take place and the tours vary on a daily basis.

In contrast, supplier milk runs are planned as cyclically repeated tours on which more than one supplier is visited before bringing the loads directly or through a consolidation centre to the plant (see (Projektgruppe Standardbelieferungsformen 2008)). That means, milk runs are served similarly to bus lines, for example every four hours or every Monday, Wednesday, and Tuesday. The underlying idea of this kind of regularity is to reduce the variability of the processes and transport schedules in the inbound network. As a result, the suppliers can reduce their safety stocks which should lead to reduced prices for parts in the long run. Further, the logistics providers can use the reliability to reduce the proportion of empty trips, and lower shipping rates can be arranged in the medium-term.

Further advantages of the milk run concept are that it allows to ship parts more frequently and in smaller lots than in case of point-to-point traffic. Compared to area forwarding, the transport times are shorter while the processes and the cost for the receiving plant are transparent. The transparency and the learning effects lead to an improved process quality.

The German Association of the Automotive Industry (VDA) defines the three above mentioned concepts – point-to-point transports, area forwarding services, and milk runs – as standard for supply by road and provides simple rules to assign the right concept. The use of milk runs, for example, is recommended for regular and stable shipments from geographically clustered suppliers whose loads are smaller than a truck load (see (Projektgruppe Standardbelieferungsformen 2008)). However, in spite of the definition as standard, and in spite of the reported advantages, milk runs do not play a significant role in Europe for inbound freight (see (Queiser 2007)).

From our perspective, two main reasons can be identified: In Europe, milk runs compete with area forwarding and point-to-point concepts not only in terms of operational cost, but mainly in terms of operational complexity. Point-to-point traffic and area forwarding services are operationally extremely easy to handle by applying rule based approaches. Due to missing decision support systems, a lot of manual effort is necessary to design and control a cost-efficient milk run system. Furthermore, milk runs are often considered only for relations to nearby suppliers served on a daily basis or more often – as it is usual in Japan. However, the geographic distribution of suppliers in Europe suggests to establish a mix of point-to-point transports, area forwarding services, and weekly recurring milk run systems.

Thus, for the successful application of milk runs to a broader range of frequent supplier plant relations three prerequisites must be fulfilled: (1) Methods from operations research must be provided, integrating and automatising the complex planning task of assigning a transport concept and scheduling the corresponding milk run tours – also for a weekly cycle time. (2) Methods to evaluate the potential of milk runs for an inbound network based on easily accessible data must be available. (3) Indicators assessing the performance of an established milk run system in relation to the alternative concepts must be introduced.

From a practical point of view, the goal of this thesis is to provide models and indicators to make the additional planning complexity for a receiving plant manageable in a way that milk runs serving the regular relations[1] can be established, or their share can be increased. The research related questions guiding the work of this thesis are presented in the following section.

[1] Please note, that the term relation in the following refers to supplier plant relation.

1.1 Research Questions

Classically, milk runs are associated with lean manufacturing. But regular transport schedules – not necessarily referred to as milk run systems – are also applied in other areas, for example the food retail sector. The literature addressing milk run and related planning problems is usually tailored to individual use cases or to the requirements of a specific sector. Hence, the first set of research questions is the following:

1. Which requirements should a milk run schedule meet? What are desirable features of a milk run schedule?

Abstracting from concrete use cases we give an overview of requirements a milk run schedule must fulfill in order to be in line with the determining factors of a supply network in the automotive industry and, more specifically, with lean manufacturing principles. Based on this, we also identify additional desirable features.

The experience shows that none of the three standard concepts defined by the VDA (Projektgruppe Standardbelieferungsformen 2008) dominates the others in all relevant dimensions or for all relevant use cases. Hence, the relationship between the concepts should be analysed answering the following two questions:

2. How is the relationship of milk runs to other transport concepts? How do they interact?

The insights from discussing the first two sets of research questions are brought together in the definition of the milk run design problem. Based on this definition, we formalise the problem by means of mathematical optimization models guided by the following questions:

3. What are appropriate optimization models covering different relevant requirements and desirable features? What are the limits? How do different model configurations behave?

Since there exist diverse approaches addressing different aspects of the milk run design problem, we give an overview and classification of the existing literature, propose different model configurations and test their behaviour on artificial instances.

Even if methods from operations research can help to make the planning complexity of a milk run design decision manageable, it is associated with effort and cost for a producing company to establish and operate the system. Therefore, the last set of research questions is related to the issue of how the cost savings potential of implementing a milk run system can be estimated, and how the performance can be measured. That is, we try to answer the following questions:

4. Which data are necessary for evaluating milk runs as an alternative transport concept by means of the proposed models? Does there exist a "quick check" that is an easy to calculate indicator for measuring the potential of milk runs? What is a good indicator to measure the performance of an established milk run system?

In order to provide reliable results for an application in practice, this last topic is investigated based on results of a real-world use case of Bosch, a first tier automotive supplier.

1.2 Outline

From the research questions introduced in the preceding section we derived the outline of this thesis illustrated in Figure 1.1.

Chapter 2 is dedicated to describing milk runs and typical road transport concepts in the automotive industry. Since there exist no unambiguous definitions, we propose a classification scheme for road transport concepts, discuss the advantages and disadvantages, and define the term milk run as we understand it for this thesis. In *Chapter 3* we discuss the milk run concept and supplier kanban systems in the context of lean manufacturing principles. Based on these results, *Chapter 4* is dedicated to defining the integrated milk run design decision.

In *Chapter 5* we give an extensive overview of related optimization approaches proposed in literature and classify them. The assessment of the models with regard to important requirements and features of the milk run design problem is carried out in *Chapter 6*. We furthermore identify the research gaps and point out the contribution and the proceeding of the remainder of this thesis. *Chapters 7* to *9* are dedicated to proposing different models addressing the milk run design problem in an integrated fashion. We test and analyse a variety of model configurations, each focusing on a different aspect using small, artificial instances solved to optimality by means of a mixed integer programming solver.

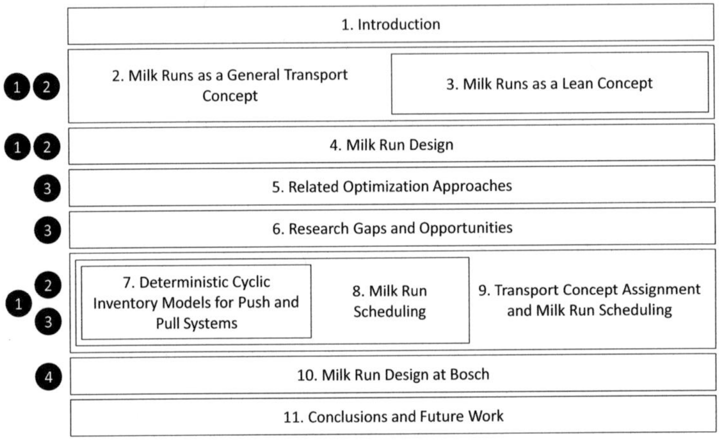

Figure 1.1: Overview of the chapters of this thesis with grey circles representing the research questions addressed

In *Chapter 10* we apply the models to the German supply network of a production site of our project partner Bosch, which is one of the biggest automotive suppliers worldwide. Based on easy-to-gather data we determine the potential of establishing a milk run system for the considered plant by solving one of the introduced model configurations for a set of resulting instances to optimality. Furthermore, we use the results to determine indicators for the "quick check" of the potential and for continuously measuring the performance of an established milk run system.

In *Chapter 11* we draw conclusions with regard to the contributions of this thesis and the results of the computational experiments, and we give an outlook on further work.

2 Milk Runs as a General Transport Concept

Milk runs are a transport concept often mentioned in the context of lean manufacturing. However, since the milk run concept is also applied by companies not explicitly following lean principles, we first introduce the general idea. In Chapter 3, we describe the logic of milk runs against the background of lean manufacturing principles.

For the definition of the milk run concept we follow Baudin (2004) saying that milk runs are "pickups and deliveries at fixed times along fixed routes" (Baudin 2004). This definition emphasizes the regularity as a key characteristic of the transport concept. In slightly different forms this concept applies to inbound, outbound and in-plant logistics. Especially in the in-plant use case, the similarities to bus lines being served at fixed intervals on regular routes become obvious: The stops at the production lines are often posted by bus stop signs along with the corresponding time table (see (Baudin 2004)).

Within this thesis we focus on the inbound use case. The outbound use case is very similar to the inbound case and, hence, all concepts introduced in the following can be applied in a very similar way. The in-plant use case is well-studied and differs from the two other ones from a planning point of view: Routing aspects usually play a minor role, and cost structures are completely different. A recent overview of the use case is given in (Kilic et al. 2012).

However, the term milk run both in literature and practice is often used ambiguously – even within the same company. The reasons for this ambiguity are the differing and heterogeneous characteristics accredited to milk runs. Thus, a more detailed definition is necessary against the background of general road transport concepts: For a better understanding of the advantages and disadvantages of milk runs we discuss general concepts for road transport common in practice and define the term milk run as we understand it for this thesis afterwards. Then we describe use cases of regular transport concepts and extract their general consistency features. We close this chapter by a short summary.

2.1 Discussion of Road Transport Concepts

We start the discussion of road transport concepts by introducing standard road transport concepts proposed in a recommendation of the German Association of the Automotive Industry (VDA) (Projektgruppe Standardbelieferungsformen 2008) for the in-bound case in more depth and shortly discuss them. Since these definitions are not enough to clearly outline the characteristics of milk runs, we furthermore propose a classification scheme for common road transport concepts followed by a discussion of advantages and disadvantages.

2.1.1 Standard Transport Concepts based on VDA

As already mentioned in Chapter 1, the VDA defines in its recommendations three standard road transport concepts from the perspective of a producing company, which is the consignee in an inbound network: direct transport or point-to-point transports, groupage service by area contract freight forwarders, and milk runs (Projektgruppe Standardbelieferungsformen 2008). The VDA does not consider specialized transports, such as bulk or silo, and Courier, Express and Parcel services (CEP), because these transport types play a minor role in the manufacturing industry.

In the following we assume that the company has no own fleet and, hence, subcontracts freight forwarders for all types of transports. Krajewska (2008) shows in a detailed analysis of trends in the freight forwarding market, that this corresponds to the typical situation for producing companies in Europe.

Direct Transport / Point-to-point Transport The simplest standard transport concept of the VDA is the direct transport with a point-to-point relation. In this case, the consignee subcontracts a freight forwarder picking up parts at a supplier and bringing them directly to the receiving plant. This transport concept is recommended by the VDA for full truck loads (FTL), which are defined according to the VDA as loads greater than 11 loading metres. The freight forwarder or logistics service provider is usually informed "today for tomorrow", that means, the freight forwarder receives the order one day before the planned pick-up date. The contractually agreed run time of the transport depends on the distance.

A common variation of this concept in practice is to consolidate Less Than Truck Loads (LTL) – loads less than 11 loading metres – of a small number of neighbouring suppliers and transport them directly to the receiving plant. In that case, the consignments of the suppliers are picked up and brought without transshipment to the receiving plant. However, this version of a direct service is not considered as a standard by VDA.

The cost for contracting a freight forwarder to operate a direct FTL transport or a direct LTL tour mainly depends on the driving distance and the tour duration. It also depends on the market situation for the freight forwarder within the source area and destination area, but it usually does not – or only to a small extent – depend on the weight or the volume.

Groupage Service by Area Contract Freight Forwarders The VDA defines groupage services operated by an area contract freight forwarder as the second standard transport concept. A groupage service is offered by a logistics service provider or by a coalition of smaller freight forwarding companies for relations from specified countries to specified countries or from specified regions to specified regions. The logistics providers run a network of consolidation centres and plan and organise the transports between the suppliers and the consignees.

The provider collects goods from suppliers on so called preliminary leg tours and consolidates them to bigger loads in the consolidation centres. On the main-leg, the consolidated loads are transported either directly to the supplier or to another consolidation centre from where the loads are deconsolidated and brought to different consignees. If the latter step exists, it is called subsequent leg. If the shipping volumes of one or two nearby suppliers already fill up the capacity of a vehicle on the pre-leg, the logistics provider conducts a direct transport from the supplier to the consignee (see for example (Schöneberg et al. 2010)).

However, it is important to understand that the responsibility to plan and to consolidate and, hence, form cost-efficient tours is given to the logistics provider. The transport price offered to the client is usually represented in so called tariff tables: In the simplest case, the price for a shipment depends on the weight and the distance from the supplier to the receiving plant. The price is independent from the actual tour executed by the freight forwarder.

Big automotive companies often subcontract for a period of one or two years different freight forwarders, which offer groupage services from and to a certain area at fixed tariffs. These providers are referred to as area contract freight forwarders, and the groupage service executed by such a provider is called area forwarding (AF) (see (Schöneberg et al. 2010)).

Usually, manufacturing companies subcontract different providers for distinct areas by conducting a bidding process. Beyond others the reason is the differing cost structure within a certain area: The capability for a freight forwarder to build cost efficient tours – filling up the loading capacity of the vehicles and the daily maximum driving time of a driver – depends to a great extent on the client structure within the area. This is reflected in the tariff tables offered to the clients.

The assignment of orders for the AF networks is also done "today for tomorrow": This means, the logistics provider and the suppliers are informed on day A (today), the pick up is performed on day B (tomorrow) and the delivery on day C – also called ABC transport in practice. For longer distances the run time of the transport might also be 48 or 72 hours instead of 24 hours. Usually, the transport volumes for a certain period are within contractually agreed limits.

Milk runs Milk runs are the third standard transport concept. Besides the regularity aspect of milk runs – fixed routes with fixed volumes – the authors of the VDA recommendation emphasise that on milk runs more than one supplier is visited before the vehicle travels directly or with transshipment to the receiving plant.

Summarizing the literature – academic work and manuals or text books for logistics practitioners – the characteristics mostly accredited to milk runs are the following:

- Milk runs are fixed routes with fixed pick up and delivery time slots (and fixed volumes) which are executed according to a fixed cycle.

- On milk runs, LTL of more than one supplier are consolidated in order to be able to increase the individual shipping frequency compared to executing a direct point-to-point transport for each of them.
- Milk runs are round tours on which full and empty returnable containers are exchanged in a ratio of 1:1.
- Milk runs are planned by the consignee.

These characteristics are partly mentioned together, often with an emphasis on one or the other. Branke et al. (2007) and Böhle and Dangelmaier (2009), for example, use the term milk run to describe the consolidation of more than one supplier to a dedicated pick up tour planned by the consignee. They do not associate any type of regularity to milk runs and do not refer to the exchange of containers or loading units.

Klug (2010a), for example, emphasizes the fact that in case of 1:1 exchanges of full and empty returnable containers the term milk run is used. He further differentiates between static and dynamic milk runs: According to his definition, static milk runs are fixed routes with more or less fixed delivery volumes executed in fixed cycles. In case of dynamic milk runs, the pickup cycles, the suppliers on a tour and/or the volumes change dynamically.

In order to be able to clearly distinct these different concepts labelled by the term milk run, we propose a set of features describing all transport concepts mentioned in this section. As before, we do not address special transports, such as bulk transports and parcel services.

2.1.2 A Classification Scheme for Transport Concepts

In their textbook about supply chain management Chopra and Meindl (2013) differentiate between different networks including direct shipping, direct shipping with milk runs, shipments via a cross dock, shipments via a cross dock using milk runs, and tailored networks being a mixture of the above. In a review on the part logistics of the German automotive industry Boysen et al. (2015) refer to these networks as point-to-point networks, direct milk run systems, cross docking systems, and a mixture of all three.

However, we think that a classification from the perspective of a manufacturing company must – besides the physical network level – include further aspects, such as the fact, if the consignee takes over tour planning tasks or outsources this task, the degree of consolidation possibilities for the logistics provider (see (Miemczyk and Holweg 2004)), or the tariff systems.

In Figure 2.1 we show the impact of the different characteristics for the area forwarding concept: In terms of tariffs the transport from the perspective of the consignee is a point-to-point connection since the costs do only depend on the distance and the weight, but not on the actual tour. On the physical level, in this example, we have a direct tour without any transshipments. The consolidation and tour planning task is completely executed by the freight forwarder.

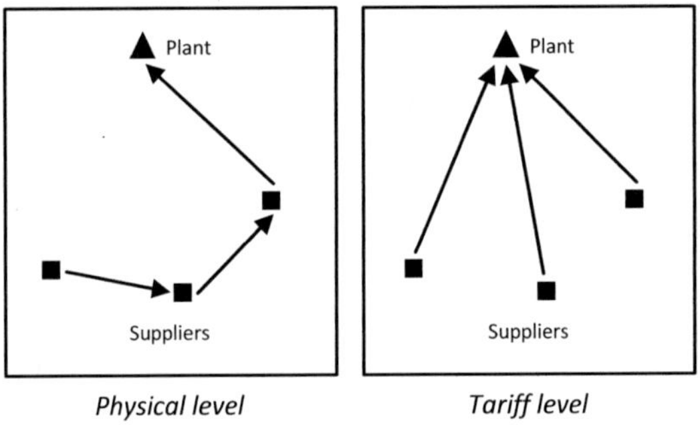

Figure 2.1: Example for the differentiation between the level of physical flows and the level of tariffs

Operational advantages obviously emerge on the physical level, while the costs are influenced on the tariff level and by the question if tour planning tasks must be executed by the consignee. Hence, in the following we classify the most typical tariff systems and the most typical forms of physical transports.

Tariff Systems

There exist many ways how transport services are invoiced by freight forwarders. However, the basic classification of different types of fulfilment cost introduced by Krajewska (2008) for the perspective of freight forwarders can be similarly applied to the cost structures of producing companies or consignees.

Cost Structure Figure 2.2 shows an overview of different types of cost for transports a consignee without an own fleet faces. In case (a), the cost depends on the driving distance and/or the duration of a tour. Often also a fixed amount is invoiced. In case (b), a vehicle is paid for example for a whole day regardless of the transports executed within that period. As already mentioned in Section 2.1.1, type (a) is typical for direct services. A payment of vehicles on a daily basis is possible, but more common in contracts between freight forwarders and their subcontractors. In both cases, the risk for inefficient, i.e. poorly utilized vehicles, is carried by the consignee.

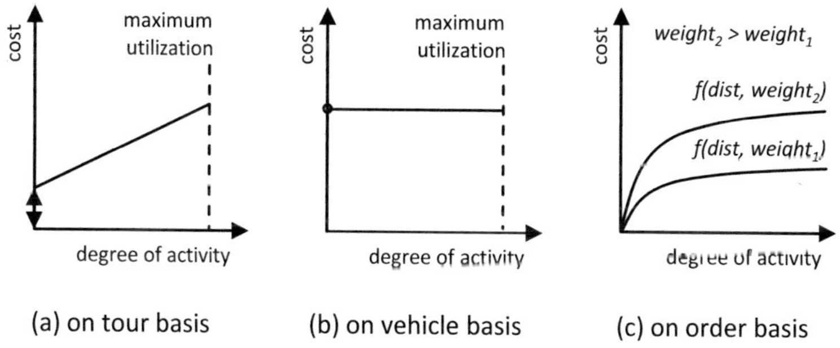

Figure 2.2: Different cost structures of transport services from the perspective of the consignee (based on a figure for the perspective of freight forwarders in (Krajewska 2008))

The payment on order basis (c) is typical for groupage services. As already mentioned in Section 2.1.1, the transport prices offered by the freight forwarder to the client is usually given in so called tariff tables considering the distance and the weight class. As shown in the figure, the cost per unit decreases with increasing weight and with increasing distance. This degressive cost structure incentivizes the consolidation of orders to bigger transport lots. However, the risk of inefficient tours is carried by the freight forwarder.

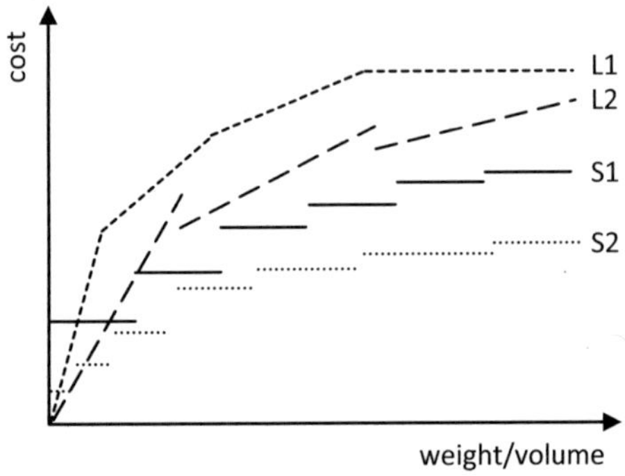

Figure 2.3: Different tariff structures for transports on order basis from the perspective of the consignee (based on (Kempkes et al. 2010))

Figure 2.3 further details the transport tariffs charged on order basis (c) for a given relation or distance class based on the work of (Kempkes et al. 2010): The cost can be either given as a piecewise linear function over weight (L1), volume, pallets, etc. If additional discounts on weight are granted, the function contains overlapping pieces in terms of the prices (L2). Furthermore, the tariffs can be given as constant prices valid within a certain weight class. These weight classes can be either of equal size (S1) or of variable size (S2). Similar examples for tariff structures are discussed in (Harks et al. 2014).

A common practice for charging on order basis is to arrange two independent tariffs for the pre-leg and the main-leg (see for example (Schöneberg et al. 2010)). Such a structure incentivises that the consignee consolidates loads from more than one supplier to the same destination on the main-leg. Furthermore, there often exist agreements on additional cost for handling in consolidation centres based on the volume or the number of consignments.

A more detailed description of the two most common cost structures – (a) and (c) – along with concrete values for a use case follows in Chapter 9 and Chapter 10.

Type of Contract As described in (Krajewska 2008) for the perspective of logistics service providers, one can divide the subcontractors of the consignee into partner freight forwarders employed on a medium-term or long-term basis and freight forwarders engaged on short-term for singular transports. As described in Section 2.1.1, these medium- or long-term contracts – for example an area forwarding contract – are agreed upon for one or two years. The tariffs are usually better compared to ad hoc arranged prices since the subcontractor has a high planning certainty, for example by arranging minimum volumes.

If a medium-term contract for regular transports, such as milk runs, is arranged, the cost structure occurring for the short-term planning task might be similar to case (b) or it corresponds to (a) with a high fixed cost share: The cost for the tour or at least a fixed part of it arises for reserving the vehicle irrespectively of whether the milk run is served or not.

Physical Transport and Tour Planning

For classifying the physical transport and the allocation of tour planning tasks between consignee and freight forwarder, we propose four features. They emphasize the differences of the concepts, but they are not necessarily independent from each other.

- **Tour Planning by Consignee:** The first feature refers to the planning activity and skills of the consignee: There are transport concepts, which require the consignee to conduct planning steps to construct concrete tours or routes through a logistics network.
- **Degree of Consolidation:** This aspect describes if a transport is exclusively dedicated to orders of the consignee or if the freight forwarder is allowed to form mixed tours with third party shipments

in order to increase cost efficiency (cmp. (Miemczyk and Holweg 2004)).

- **Direct Service:** The feature characterises the tours itself: It defines if there are further transshipments between the pick up and the delivery of parts for achieving a higher load factor by consolidation (and deconsolidation). Note that there might be transshipments for other reasons than consolidation – for example multi-modal transports. However, we do not consider them in our work.
- **Planning horizon / Regularity:** The last aspect refers to the planning and communication horizon of the consignee: It states if the consignee plans and communicates transports ad hoc on a short-term, that means, in the extreme case "today for tomorrow" or if the consignee communicates/arranges regular transport plans in order to increase the predictability for his freight forwarders and the suppliers.

Figure 2.4 gives a simplified overview of the features characterising a transport concept: The physical transport and tour planning features define the transport concept, while the applied cost structure and type of contract are a consequence of the transport concepts and, at the same time, characterise them.

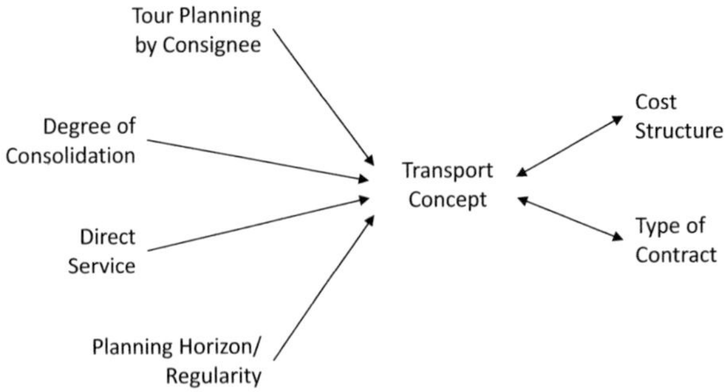

Figure 2.4: Features and aspects characterising a transport concept

In the following, we shortly describe all transport concepts resulting from enumerating the sensible combinations of the physical transport and tour planning features. Furthermore, for each concept the typical tariff structures and types of contracts are given. For a better overview, we split the descriptions by the feature "planning horizon / regularity" into ad hoc transport concepts and transport concepts with regular components.

Transport concepts with ad hoc planning Figure 2.5 shows all transport concepts resulting from enumerating the reasonable combinations of the features for the ad hoc planning case.

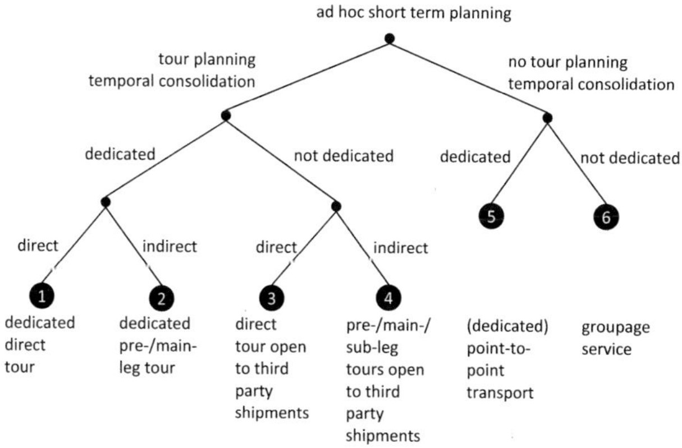

Figure 2.5: Ad hoc transport concepts resulting from enumerating the sensible combinations of the features from the perspective of a (single) consignee

These concepts can be briefly described as follows:

1. **Dedicated direct tour:** Consignee plans direct tours for its own suppliers on short-term and subcontracts freight forwarders for the execution of the tours ad hoc with ad hoc tariffs or tariffs based on medium-term contracts.
2. **Dedicated pre- and main-leg tour:** Consignee plans pre- and main-leg-tours for its own suppliers and subcontracts freight forwarders for the execution of the tours and the transshipments in

the consolidation centres of the freight forwarder (or its own plants) with ad hoc tariffs or tariffs based on medium-term contracts.

3. **Direct tour open to 3rd party**: Consignee plans direct tours for its own suppliers and subcontracts a freight forwarder ad hoc with ad hoc tariffs or tariffs based on medium-term contracts for the execution of the tour but allows to add 3rd party shipments or allows to deconsolidate the tour and build new, more efficient ones with 3rd party shipments.
4. **Pre- and Main-leg-tours open to 3rd party:** Consignee plans pre- and main-leg tours for its own suppliers and subtracts a freight forwarder ad hoc with ad hoc tariffs or tariffs based on medium-term contracts for the execution of the tours and the transshipment but allows to add 3rd party shipments or allows to deconsolidate the tour and build new, more efficient ones with 3rd party shipments.
5. **Dedicated point-to-point transport:** Consignee schedules dedicated point-to-point transports – using as good as possible the vehicle capacity – and subcontracts a freight forwarder for the execution of the transport ad hoc with ad hoc tariffs or tariffs based on medium-term contracts.
6. **Groupage service:** Consignee outsources individual shipments to a freight forwarder paying fixed (distance and weight based) tariffs mostly based on medium-term contracts. The freight forwarder plans and executes the transport using its consolidation centres and builds tours including 3rd party shipments.

The "on tour basis" cost structure is predominant for concepts 1 to 5, while groupage services are charged on order basis. To plan ad hoc direct or indirect tours and leave them completely open to changes by the freight forwarder has no advantage compared to the groupage service from an operational point of view. This concept is only beneficial for negotiating better rates than in a groupage service since the tour cost give an upper bound for the cost incurred when paying on order basis.

Table 2.1 gives a detailed overview of the advantages and disadvantages for every single transport concept based on the following criteria: planning complexity for the consignee, transparency – assuming that transparent processes are more reliable on the long run –, transparency of the

Table 2.1: Advantages (+) and disadvantages (–) of ad hoc transport concepts

Advantages	Disadvantages
1 Dedicated direct tour	
+ possibly cheaper than groupage service + short run times + complete cost and process transparency + appropriate for all shipment sizes	– high daily planning complexity – risk of inefficient tours
2 Dedicated pre- and main-leg tour	
+ possibly cheaper than dedicated direct tours and groupage service + almost complete cost and process transparency + appropriate for all shipment sizes	– very high daily planning complexity – necessity of knowledge on operational capacity and processes of consolidation centres – risk of inefficient tours
3 Direct tour open to 3rd party	
+ possibly cheaper than dedicated direct tours and groupage service + short run times + partial cost transparency + appropriate for all shipment sizes	– high daily planning complexity – no process transparency – potentially complex pricing
4 Pre- and Main-leg-tours open to 3rd party	
+ possibly cheaper than direct tours and groupage service + partial cost transparency + appropriate for all shipment sizes	– very high daily planning complexity – knowledge on capacity and processes of consolidation ctr. necessary – no process transparency – potentially complex pricing
5 Dedicated point-to-point transport	
+ lowest transport cost + reliable processes + complete cost and process transparency + short run times + easy to plan and control + stable and easy to calculate cost	– appropriate only for suppliers with high volumes, otherwise the big transport lot sizes lead to high inventories
6 Groupage service	
+ easy to plan and control + risk for inefficient tours is outsourced + appropriate for all shipment sizes + stable and easy to calculate cost	– probably higher transport cost – no cost and process transparency – 3rd party shipments might lead to higher risk of delays

cost, risk of building inefficient transports, the applicability to different load sizes (FTL or LTL) and the costs occurring from the respective concept. A discussion of the corresponding advantages and disadvantages over all transport concepts follows in the subsequent subsection.

Transport concepts with regular components Figure 2.6 shows all concepts resulting if the consignee plans in the medium-term – for example for several weeks or months – in order to achieve a certain level of regularity.

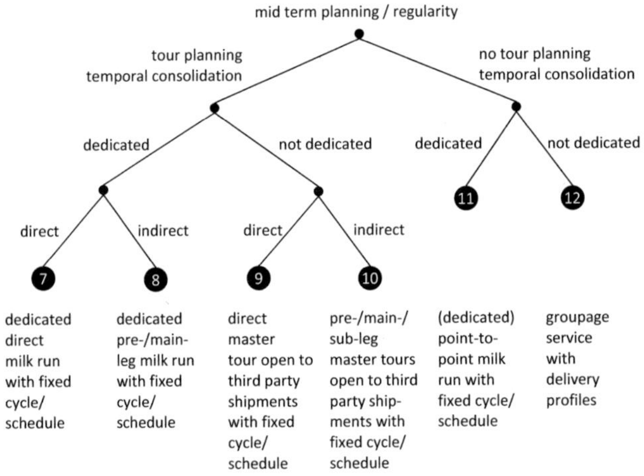

Figure 2.6: Transport concepts with regular components resulting from enumerating the sensible combinations of the features from the perspective of a (single) consignee

We can briefly describe the resulting concepts as follows:

7. **Dedicated direct milk run with fixed cycle or schedule:** Consignee plans regularly (i.e. daily or weekly) recurring dedicated direct tours and subcontracts freight forwarders to execute these tours for an agreed period.
8. **Dedicated pre- and main-leg milk run with fixed cycle or schedule:** Consignee plans regularly (i.e. daily or weekly) recur-

ring dedicated pre- and main-leg-tours and subcontracts freight forwarders to execute these tours and the transshipments in the consolidation centres for an agreed period.

9. **Direct master tour open to 3rd party shipments with fixed cycle or schedule:** Consignee plans regularly (i.e. daily or weekly) recurring direct tours and subcontracts freight forwarders to execute these tours for an agreed period. The freight forwarder is allowed to add 3rd party shipments – or ad hoc shipments of the consignee. It might be agreed that the freight forwarder deconsolidates the tours and builds new, more efficient ones with 3rd party shipments on a daily basis.
10. **Pre-, Main- and Sub-leg master tour open to 3rd party shipments with fixed cycle or schedule:** Consignee plans regularly (i.e. daily or weekly) recurring pre- and main-leg- tours and subcontracts freight forwarders to execute these tours and the transshipments for an agreed period. The freight forwarder is allowed to add 3rd party shipments and, probably an additional transshipment along with a sub-leg tour. As in concept 9 the possibility to deconsolidate tours might be arranged.
11. **(Dedicated) point-to-point milk run with fixed cycle or schedule:** Consignee plans regularly (i.e. daily or weekly) recurring point-to-point transports tours and subcontracts freight forwarders to execute these tours for an agreed period.
12. **Groupage service with delivery profiles:** Consignee arranges regular delivery profiles or cycles (i.e. Mon/Wed/Fri or every second working day) for the groupage service paying fixed (distance and weight based) tariffs. The freight forwarder plans and executes the transports using its consolidation centres and building tours including 3rd party shipments.

As for the ad hoc case, concepts 7 to 11 are usually invoiced on tour basis, while the groupage service is invoiced on order basis. Compared to tariffs for ad hoc concepts, the tariffs here should be lower due to the planning certainty for the freight forwarder.

Table 2.2: Advantages (+) and disadvantages (–) of transport concepts with regular components (part 1 of 2)

Advantages	Disadvantages
7 Dedicated direct milk run with fixed cycle or schedule	
+ possibly cheaper than ad hoc concepts due to lower risk for freight forwarder and possibly cheaper than groupage service with delivery profiles + short run times + complete cost and process transparency + high reliability due to learning effects through repetitive processes + improved planning ability for consignee, suppliers and freight forwarders due to fixed dates and arrival times + appropriate for all shipment sizes	– high complexity for tactical planning – appropriate only for suppliers with stable frequencies and approximately stable volumes – necessity of adapting order policies – risk of inefficient tours
8 Dedicated pre-/main-leg-milk run with fixed cycle or schedule	
+ possibly cheaper than dedicated direct milk runs + almost complete cost and process transparency + high reliability due to learning effects through repetitive processes + improved planning ability for suppliers and freight forwarders due to fixed dates and time for pick ups + appropriate for all shipment sizes	– high complexity for tactical planning – appropriate only for suppliers with stable frequencies and approximately stable volumes – necessity of adapting order policies – risk of inefficient tours
9 Direct master tour open to 3rd party shipments with fixed cycle or schedule	
+ possibly cheaper than dedicated direct milk runs + lower risk of inefficient tours because the freight forwarder has time to acquire additional load + short run times + partial cost and process transparency + partially higher reliability due to learning effects + improved planning ability for consignee, suppliers and freight forwarders due to fixed dates and approximately reliable arrival times + appropriate for all shipment sizes	– high complexity for tactical planning – appropriate only for suppliers with stable frequencies and approximately stable volumes – necessity of adapting order policies – potentially complex pricing

Table 2.2: Advantages (+) and disadvantages (–) of transport concepts with regular components (part 2 of 2)

Advantages	Disadvantages
10 Pre-/Main-/Sub-leg master tour open to 3rd party with fixed cycle or schedule	
+ possibly cheaper than direct master tours + lower risk of inefficient tours because the freight forwarder has time to acquire additional load + partial cost and process transparency + partially higher reliability due to learning effects through repetitive processes + improved planning ability for consignee, suppliers and freight forwarders due to fixed dates and approximately reliable arrival times + appropriate for all shipment sizes	– high complexity for tactical planning – appropriate only for suppliers with stable frequencies and approximately stable volumes – necessity of adapting order policies – potentially complex pricing
11 (Dedicated) point-to-point milk run with fixed cycle or schedule	
+ cheapest regular transport concept + short run times + complete cost and process transparency + high reliability due to learning effects through repetitive processes + improved planning ability for consignee, suppliers and freight forwarders due to fixed dates and arrival times	– appropriate only for suppliers with high volumes, otherwise the big transport lot size leads to high inventories – higher cost than purely demand driven ad hoc point-to-point transport – necessity of adapting order policies
12 Groupage service with delivery profiles	
+ easy to plan and control + risk for inefficient tours is outsourced + appropriate for all shipment sizes + stable and easy to calculate transport cost + improved capacity planning ability for consignee, suppliers and freight forwarders due to fixed dates (but not times)	– probably higher transport cost – no cost and process transparency – 3rd party shipments might lead to higher risk of delays – necessity of adapting order policies

All concepts have in common that a fixed cycle or schedule is considered – either on the level of tours or on the level of shipments. The degree of regularity varies in the sense that either dates and arrival times are fixed or just the dates.

Compared to the ad hoc concept 3 and 4, the master tours open to 3rd party shipments – concepts 9 and 10 – provide advantages on the operational level to the consignee: The suppliers know the day and possibly the approximate arrival times in advance. The freight forwarder has planning certainty and, hence, a higher chance to acquire additional freight. The master tour concept is similar to the so called dynamic milk run described in (Klug 2010a). However, for a better distinction to the milk run concept, we use the term master tour. A similar definition for master routes is given in (Lourenço and Ribeiro 2011) and (Ratliff and Nulty 1997) saying that the concept consists of defining the master routes and afterwards adjusting them on a daily basis depending on the demand and the vehicle capacity. These definitions do not specify if a subcontractor or the consignee performs the daily planning tasks.

In (Schöneberg et al. 2010), (Schöneberg et al. 2013) and (Meyer et al. 2011) the concept of delivery profiles for area forwarding is described corresponding to our concept 12: The delivery profiles combine the advantages of groupage service with the advantages of regular shipment days known from milk runs or master tours. A validity period of 3 months for the delivery profiles is proposed in (Schöneberg et al. 2010).

All regular concepts, obviously, have in common that they are only applicable for suppliers with stable shipping frequencies and approximately stable volumes. A detailed list of advantages and disadvantages for every single concept is given in Tables 2.2 and 2.2. A general discussion follows in the following section.

2.1.3 Advantages and Disadvantages

In general one can state that for planning tours – direct or even indirect ones – the consignee needs skilled people and, ideally, advanced decision support systems (see right hand side of Figure 2.7). A daily operational planning of ad hoc tours, obviously, causes a bigger effort than a tactical planning of regular milk runs or master tours. However, only if tours are planned by the consignee – and hence orders are consolidated to tours – transport costs become transparent, or at least an upper bound can be determined.

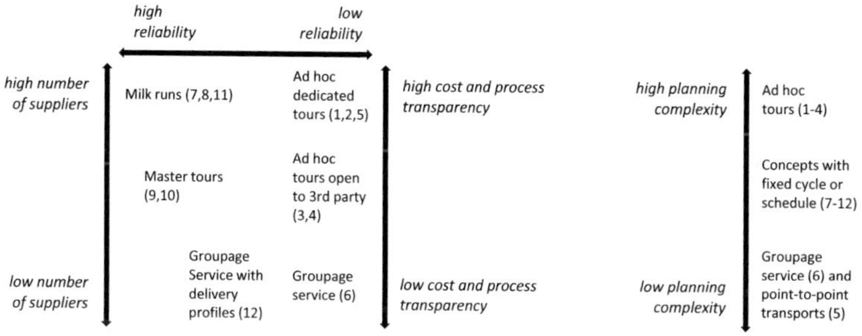

Figure 2.7: Simplified overview of characteristics (italic letters) of road transport concepts for less than truck loads

In terms of reliability all concepts with regular cycles or schedules are beneficial (see left hand side of Figure 2.7): If at least the dates for pick ups are known in advance, suppliers can control their production accordingly and decrease the stock of finished goods. The freight forwarder has a better planning certainty for the vehicle and transshipment capacity needed and needs fewer or less expensive ad hoc rented vehicles. Furthermore, the freight forwarder has a higher opportunity of acquiring additional 3rd party freight to fill up tours or to avoid empty return trips to the depot. This should lead to better tariffs for tours as a whole and for orders sent by groupage service. If in case of milk runs also the arrival times are fixed, the capacities in the respective shipping areas and consolidation centres can be better planned. Furthermore, the regularity not only leads to a better capacity usage but also allows to exploit learning effects improving the process quality and decreasing the process times.

Dedicated tours are in both cases, ad hoc tours and regular tours, more reliable (see left hand side of Figure 2.7) since the consignee can control and influence the process quality of his own suppliers and the freight forwarder. Whenever 3rd party shipments are transported on the same truck, it becomes more difficult to control the performance and to improve it.

On the other hand, considering 3rd party shipments can reduce the cost tremendously if the consignee has only a small number of suppliers with small shipment sizes in general or in a specific area (see left hand side of Figure 2.7). This results from the effect that the risk for inefficient tours declines the more clients are served. Or the other way around: The higher the number of clients, the better is the optimization potential which leads to a better capacity usage and, hence, to lower average cost for a single shipment. Furthermore, there occur "common" effects of economies of scale, for example for the consolidation centres. On the other hand, a freight forwarder hedges against the risk of expensive tours by adding a margin for risk and, of course, for profit and the actual cost are obscured to the consignee.

Building point-to-point transports with the loads of one supplier filling up the vehicle capacity is by far the cheapest transport concept. However, it is only applicable for parts with a high demand in terms of volume or mass. Otherwise the big transport lots lead to high inventories and, most probably, to high cost.

2.1.4 Implicit Outsourcing Decisions and Supply Chain Coordination

From the perspective of a consignee, the selection of a cost efficient transport concept for a supplier plant relation is complex: It cannot be taken separately for every relation because the costs for some concepts depend on the consolidation of suppliers to a route and the concrete sequence in which they are served.

However, a second source of complexity is that factors which are relevant to the decision are hard to evaluate in monetary terms: It is, for example, difficult to assess the value of reduced planning complexity and reduced risk for inefficient tours compared to losing the control over costs and processes. Furthermore, it is hard to assess how big the benefits of reliable schedules are for the consignee, the suppliers and the logistics service providers.

The first example addresses typical aspects treated in literature considering outsourcing decisions, while the second aspect is treated under the term supply chain coordination.

Outsourcing Decisions

The question related to outsourcing in our case is to which extent a consignee outsources – beside the execution of the transports – the planning process for the transports.

Table 2.3 outlines the results of an analysis of main motives and risks of general outsourcing decisions deducted in (Quélin and Duhamel 2003). Applied to transports within inbound networks it can be said: The main motive for outsourcing is the additional optimization potential due to 3rd party customers of the freight forwarder and a better utilization of capacity (see also (Hsiao et al. 2010)) leading to reduced operational cost. Furthermore, a consignee focuses on its core competencies, has a smaller investment into software tools for planning and controlling transports including, for example, tracking and tracing for real-time disruption management. At the same time the consignee uses the external competencies of a logistics service provider. Area forwarding tariff tables provide purely variable cost scaling with increasing or decreasing volumes.

On the other hand, the dependency of the logistics service provider increases. The loss of know how even reinforces the dependency and, at the same time, decreases the ability to recognize hidden cost.

In case of outsourcing decisions it is especially difficult to profoundly evaluate the risks arising for the supply chain from outsourcing. An evaluation requires advanced methods clearly beyond the scope of this thesis. For more details on these methods we refer to (Heckmann 2015).

Furthermore, in practice the decision to transfer the tour planning process to an area forwarding network provider is already taken. Therefore, in this thesis we rather ask the question if transport concepts with regular components resulting from tactical planning by the consignee can be introduced without increasing operational cost. If the latter is possible, a part of the tour planning process can be shifted back to the consignee, not on the very complex operational level, but rather on the tactical one.

Table 2.3: Main motives and negative outcomes of outsourcing reported in literature according to an analysis of Quélin and Duhamel (2003)

Main motives for outsourcing	Main negative outcomes
+ To reduce operational costs + To focus on core competencies + To reduce capital invested + To improve measurability of costs + To gain access to external competencies and to improve quality + To transform fixed costs into variable costs + To regain control over internal departments	– Dependence on the supplier – Hidden costs – Loss of know-how – Service providers lack necessary capabilities – Social risk

Supply Chain Coordination

Common replenishment cycles – that means, schedules with regular components as milk runs – are proposed in literature (i.e. (Viswanathan and Piplani 2001) and (Piplani and Viswanathan 2004)) as one possible coordination mechanism for supply chains with two or more levels. There exists a wide range of literature dealing with the evaluation of positive effects of coordination mechanisms – especially with respect to inventory cost – by analytical and simulation approaches (for a recent overview see for example (Bahinipati et al. 2009) or (Chan and Chan 2010)).

In practice, these approaches are often difficult to apply since the receiving plant has no or insufficient information about production schedules, inventory levels and cost structures of the suppliers, unless a supplier kanban system is applied (see Chapter 3). Therefore, in this thesis we assume that positive coordination effects on the inventory of the suppliers exist but do not explicitly consider them. Furthermore, we assume that a freight forwarder is able to offer better tariffs due to the planning certainty if regular schedules are arranged, which leads to lower operational cost for the consignee.

In summary, we focus on the question for which supplier relations regular transport concepts might be beneficial in terms of operational cost keeping the additional advantages in mind.

2.2 Definition of Milk Runs

Based on the preceding sections, we define the milk run concept for the inbound use case as follows:

- A milk run is a concept to serve supplier relations with regular volumes.
- It is a fixed tour with a fixed sequence of stops serving at least one supplier and being executed cyclically or according to a fixed schedule.
- The volumes are determined on a daily basis by an order policy aiming for levelled load sizes or simply by aggregating the demand until the next shipment (or by a supplier kanban system, see Chapter 3).
- The milk run is planned and communicated to the suppliers by the consignee. The execution is usually outsourced to a freight forwarder.
- The milk run plan is valid for several weeks or months in order to achieve the positive effects.
- A milk run might contain a transshipment.
- The milk run can be – but does not have to be – a round tour starting and ending at the receiving plant in order to allow an exchange of full and empty returnable containers.
- The cost for a milk run usually depends on the distance and the duration of the tour. It is usually fixed in a midium-term contract and, at least, a part needs to be paid for reserving the vehicle if the milk run is not performed. Compared to the same tour, which is outsourced ad hoc, the rates should be lower.

Groupage service with delivery profiles can be seen as an alternative to milk runs for regular supplier plant relations. For the consignee it implies a lower planning complexity and no risk for inefficient tours, but the cost and processes are not transparent and it might lead to higher cost.

2.3 Use Cases of Transport Concepts with Regular Components

In order to show the importance of regular transport concepts, in this section we describe use cases along with the advantages reported by the users of the corresponding concept. The applied concepts vary from pure milk run systems to delivery profiles and are from different industrial sectors.

Peterson et al. (2010) report of a milk run system implemented by a plant of the Bosch/Siemens Home Appliances Corporation located in North Carolina. The 75 most important suppliers, which make up 80% of the total shipping volume, are served on weekly or bi-weekly scheduled direct milk runs operated by a logistics service provider. The authors emphasize two advantages of the milk runs compared to ad hoc concepts: The predictability of fixed and stable distribution patterns and the possibility to negotiate better rates.

In the Netherlands, Albert Heijn, a leading super market chain, supplies its stores with weekly recurring tours valid between 3 and 6 months (see (Gaur and Fisher 2004)). The advantages of periodic schedules reported by (Gaur and Fisher 2004) are: The workforce management for loading trucks in the distribution centres and for unloading trucks and stocking shelves at the stores is simplified. Due to the periodic schedule, the stores know when to place their orders, while the suppliers synchronise their deliveries with the schedules. This allows to reduce buffers in the distribution centres and, in the best case, to apply a pure cross docking from the inbound to the outbound trucks. Furthermore, the specific predetermined routes each day allow Albert Heijn to negotiate cheaper long-term contracts with truck leasing companies.

Ekici et al. (2014) report about cyclic delivery tour schedules applied by an industrial gas manufacturer. These cyclic deliveries are applied to simplify the combined inventory and routing problem in a vendor managed inventory (VMI) system – a system in which the vendor controls the inventory at the store of a client. Furthermore, cyclic schedules – repeated in this case every two or seven days – provide a more stable replenishment and delivery plan "desirable by the vendor and customers for planning and coordination purposes" (Ekici et al. 2014).

In the work of Spliet (2013), the advantages of fixed schedules with fixed time windows for pick ups and deliveries for a distribution network are outlined as follows: Fixed schedules allow an easy planning of packing in distribution centres, they allow the customers to roster the delivery handling personnel, and the inventory control and the sales management are simplified. Spliet (2013) cites an unpublished study about the distribution networks of four large retail chains that states that already small deviations from the delivery schedule may lead to significant cost increases. This is mainly caused by the effect that a one hour delay results on average in 5.5 additional man hours for the handling crew. The author furthermore emphasizes the positive effects if the same driver always visits the same customers, which was first discussed in literature by Groër et al. (2009). Similarly, Smilowitz et al. (2013) and Coelho et al. (2012), who do not consider a specific use case, discuss the effect of regularity or consistency of plans for improving service quality and driver efficiency in general.

The work of (Schöneberg et al. 2010) and (Schöneberg et al. 2013) considers delivery profiles for area forwarding inbound networks of large scale companies especially in the context of the automotive industry. The authors report that the main advantage of these fixed delivery profiles – or fixed delivery days – is that the stability simplifies the planning for supply chain partners and enables freight consolidation "without causing high invenstments in IT-systems integration" (Schöneberg et al. 2010): The reason is that material requirements planning systems common in practice, such as SAP, support regular delivery patterns. Literature with a focus on coordination mechanisms of supply chain systems refers to delivery profiles (Viswanathan and Piplani 2001) as "common replenishment epochs".

In (Meyer et al. 2011), delivery profiles – called weekly transportation patterns in this work – are used as a coordination mechanism to enable a logistics network provider to build weekly recurring and, hence, reliable tour plans for all of his clients in the network. It could be shown in a simulation study on real network data that the network provider can reduce the number of pre-leg and sub-leg tours considerably on the one hand and that the network loads can be levelled out over the course of the week.

2.4 Classification of Consistency Requirements

The examples of the preceding section all consider a somehow regular medium-term transport schedule. From these examples we can derive the following types of medium-term schedule regularity – or stated differently – consistency:

- **Quantity consistency:** Every shipment is approximately of the same size.
- **Frequency consistency:** The number of shipments during a certain period, i.e. one day or one week, is the same.
- **Inter arrival time consitency (ITC):** The time in between two shipments is approximately the same. That means, the visits occur in constant cycles.
- **Arrival time consisteny (ATC):** The arrival time for a pick up or a delivery is approximately the same, for example always between 14:00 h and 15:00 h.
- **Driver consistency (DC):** The driver serving a supplier is always the same, or it is just a small group of drivers serving a supplier.

2.5 Summary

Since in literature and industrial practice the term milk run is used ambiguously due to the heterogeneous characteristics accredited to this concept, we proposed a general classification scheme for common road transport concepts used in Europe. Following this scheme, we discussed advantages and disadvantages of transport concepts with regular components compared to their ad hoc counterparts and defined the milk run concept.

By gathering use cases from literature we showed that regularity in transport schedules plays an important role in different industrial areas due to three main motives: better transport tariffs, reduced operational complexity and a better planning ability for suppliers and freight forwarders.

The overview of use cases, furthermore, showed that – besides daily milk runs – weekly recurring schedules are common in industrial practice. In this context consistency requirements are proposed to further reduce the operational complexity and promote learning effects and a better relationship between drivers and suppliers.

For the decision which transport concept is the best for regular supplier plant relations, the consignee needs to consider aspects relevant to outsourcing and to supply chain coordination. However, for this thesis, we assume that the introduction of regularity improves the coordination between the consignee, the suppliers and freight forwarders without measuring the medium-term effect. Furthermore, we assume that milk runs offer a higher transparency both in cost and processes and require, at the same time, a higher planning ability on the tactical level. We do not consider explicitly these positive medium-term effects. Instead we focus on providing planning models and solution procedures to support the tactical decision and make the increased planning effort manageable for the consignee.

Point-to-point milk runs and master tours can be interpreted as a special form of milk runs, and groupage service with delivery profiles is a special case of groupage service. Hence in this thesis, we focus on the following question: For which regular supplier plant relations are milk runs beneficial in terms of direct – measurable – operational costs compared to groupage service with or without delivery profiles?

3 Milk Runs as a Lean Concept

As mentioned in Chapter 1, the need "to move small quantities of a large number of items both between and within plants with short, predictable lead times and without multiplying transportation cost" (Baudin 2004) has driven lean manufacturers to organize transports at fixed times along fixed routes, that is in the form of milk runs.

To fully understand the logic of milk runs against the background of lean manufacturing principles we describe the origins of lean thinking, introduce its basic principles and, most important for this thesis, describe milk runs for supplier kanban systems. Furthermore, we depict how heijunka levelling mechanisms can be used to reach volume consistent tours on weekly recurring milk runs.

3.1 Origins of Lean Manufacturing

The origins of lean thinking are the shop-floors of Japanese manufacturers after World War II. The reason for the development was, beyond others, a small but intense domestic competition on the car market, which asked for a great variety of car types – from luxury cars over trucks to small cars for the overcrowded cities. In contrast to the mass production in the US and Europe, in Japan the manufacturers had to produce cars in a great variety in small lots (see for example (Womack et al. 1990), (Monden 2012) or (Hines et al. 2004)).

In these early days, the engineers, especially at Toyota, primarily tried to eliminate all types of waste and excess – such as inventories, excess production or setup and waiting times – when designing the production processes. This resulted in the so called Toyota Production System, the base of lean manufacturing. Beyond other important innovations there emerged the just-in-time production system (JIT), the kanban system for implementing a pull based flow and automated mistake proofing techniques, but also management principles such as respect for employees (cmp. (Hines et al. 2004)).

Since these days, the lean concept was continuously evolved. This was partly driven by new methods such as value stream mapping. It focuses more on value creation than waste reduction and extends the lean management principles beyond manufacturing and the scope of a single company: The concept stretches "from customer needs right back to raw material sources" (Hines et al. 2004). Another driver for innovations were progressing technologies, such as reliable and cheap internet connections enabling the application of so called electronic kanban, which are much more flexible than its paper based counterparts (see Section 3.3).

Matzka et al. describe the main focus of lean manufacturing as "an all-round improvement in the economics of manufacturing operations by eliminating wastes of all forms, inventory included, in order to make the production more economical" (Matzka et al. 2012). The main principles also apply to the logistics services, which connect value-adding production processes.

3.2 Basic Principles of Lean Manufacturing

Before introducing the milk run concept within lean manufacturing, the most relevant lean concepts in its basic form are characterised. In order to do this, we mainly follow the description of the lean design principles of Rother and Shook (1999):

The *takt time* defines the number of times one part or product should be produced during a shift or a day in order to meet customer requirements. It synchronises the pace of production with the pace of sales at the pacemaker process. The production of different products or parts at the pacemaker process should be evenly distributed over time and within small lots in order to achieve a levelled production mix for the upstream processes. The pacemaker process, which is the bottleneck process within the system, is the only point, where production is scheduled based on the expected customer demand.

The principle of *continuous flow* "refers to producing at a time, with each item passed immediately from one process step to the next" (Rother and Shook 1999). Under these circumstances there is no inventory between two process steps. However, in case of a supply process, where the supply site is not right next to the place of consumption, transportation is needed and in most of the cases it is not realistic to ship one piece at a time, especially, not in case of supplier milk run systems (cmp. (Rother and Shook 1999)). However, in a flow optimized logistic the transport batches are as small as possible.

The processes which can not be connected by continuous flow, i.e. processes, at which batching is necessary, – both within the plant and from a supplier to a plant – are connected by a *pull system*: In a pull system the production is demand driven, that means, the production of a preceding process is restricted to the quantity required by its downstream customer. Stated differently, the production depends on the status of the production. In contrast, in a push system the production is scheduled in advance in accordance to a demand forecast of the downstream process (cmp. (Hopp and Spearman 2001a)). One possibility to implement a pull system between two processes are kanban loops which are introduced in the next section. The levelled pacemaker process ideally causes a levelled use of resources by the pull controlled upstream processes.

As a consequence of applying the described principles, the processes are stable and robust (see (Klug 2010b)). This is reinforced by another very important part of lean manufacturing: the *continuous effort to improve* the processes.

For a more detailed introduction to the principles of lean manufacturing and its roots – the Toyota Production System – the reader is referred to (Womack et al. 1990), (Liker 2004), (Monden 2012) or (Rother and Shook 1999).

3.3 Milk Runs and Supplier Kanban Systems

As mentioned in the preceding sections, two value adding processes in lean manufacturing are often connected by kanban loops, which is a technique developed by Toyota to implement a pull system. In the following two sections we shortly introduce the basic kanban mechanisms focussing on supplier kanban systems and milk runs.

3.3.1 Kanban Mechanisms

In a kanban system – kanban is Japanese for "tag" or "card" – only the amount of goods actually consumed by a process are re-ordered from the preceding process or source: This is achieved by attaching a kanban carrying all necessary product information to every part (or a batch of parts defined by the size of a container)[1]. When a part is consumed, the corresponding free card is detached and brought back to the source of the part. There it serves as a signal to re-produce the corresponding part or to withdraw it from a store and deliver it to the consuming process. Consequently, a kanban system can be understood as an information system controlling the daily production quantities, and it is a form of implementing the just in time (JIT) production method.

There exist two kinds of kanban mainly used: A *withdrawal* kanban specifies the type and the quantity of parts, which the succeeding process withdraws from the preceding process. A *production ordering* kanban instructs the preceding process to produce a part in the required quantity. If an external supplier is connected to a plant, this is usually realized by establishing a withdrawal kanban system. In this case, it is also called *supplier* kanban or external kanban system (see for example (Monden 2012)), and the cards cycle on regular transports between the consuming plant and a supplier providing purchased parts.

The daily information in form of kanban is also referred to as short-term call orders, specifying exactly the amount of parts needed on short-term and existing also in push based production systems (see (Boysen et al. 2015)). In lean manufacturing, besides this short-term information, the receiving plant also communicates its planned medium-term production schedule for the next one (or two) month(s) in the middle of the preceding month, and the cycle time and the number of kanban for the corresponding supplier loop. With this information the vendor can plan its own cycle times and the necessary number of kanban to its own suppliers.

Since the number of kanban is fixed in every cycle, the total amount of parts in the system and, hence, the so called Work in Progress (WIP)

[1] For a better readability, in the following, we assume that one kanban corresponds to one part.

can be controlled and kept on a low level. Hopp (2011) calls this effect the "Magic of Pull" because low levels of WIP cause short cycle times and "spurred by the pressure" (Hopp and Spearman 2001a) create high quality levels (see (Hopp and Spearman 2001a)).

Especially, in case of supply processes with remote sources, physical kanban are replaced by so called *electronic* kanban (or eKanban) (see (Monden 2012)): Depending on the implementation, the withdrawal of a part can be transferred almost in real time via an IT-network to the supply source. That means, among other advantages, a reduced fluctuation in order quantity caused by transport delays, an easier adaptation of the number of kanban and a greater efficiency in the control of kanban (a detailed discussion of the advantages of eKanban can be found in (Kotani 2007)). Furthermore, eKanban also allow for the sequenced withdrawal method: The production or loading sequence is included on the kanban information so that the supplier can deliver the parts in the right mix and ordered by the sequence of production during the next cycle. For a detailed description of the information and material flows the reader is referred to the dedicated chapters in (Monden 2012). A discussion of hybrid kanban control systems can be found, for example, in (Dickmann 2009).

Despite the improved conditions through the application of eKanban, a study of the Lepros GmbH from 2007[2] cited by Dickmann (2009) shows that supplier kanban systems are rather rare with around 10% of all types of kanban systems. Production kanban systems, for example, represent 55% and in-plant withdrawal kanban systems around 25%. Unfortunately, we do not have direct access to this study and do not know the region where it was carried out. However, the results meet our expectations derived from projects with industrial partners from the automotive industry in Germany.

3.3.2 Combining Supplier Kanban Systems and Milk Runs

Generally, there exist two types of withdrawal systems: In a *constant-quantity* withdrawal system, the parts are conveyed in constant quantities

[2] Lepros (2007): "Fakten zu Lean" Umfrage 2007, Lepros GmbH.

at varying times, whereas in the *constant-cycle* (constant interval) system, parts are transported at regular times. For supplier kanban systems always the latter type is applied. The transports are organised in round-tours, called round-tour mixed-loading system (see (Monden 2012)), milk run (see (Baudin 2004)) or milk-round (see (Bicheno and Holweg 2009)) if more than one supplier is visited on one tour.

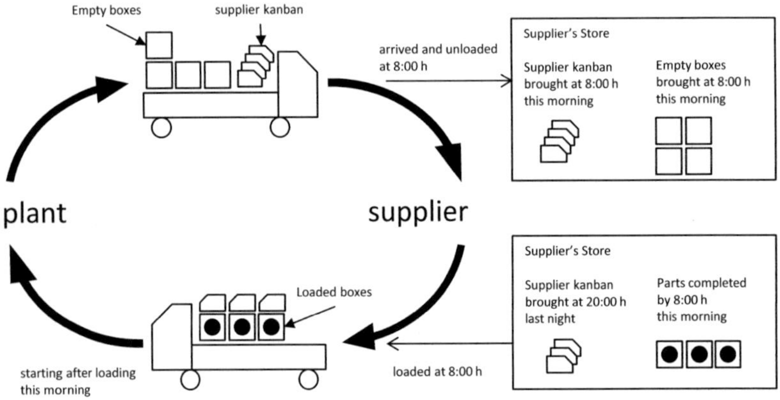

Figure 3.1: Example for a Supplier Milk run with two shipments per day with arrival times at the supplier scheduled at 8:00 h and 20:00 h and a reproduction lead time of one cycle based on an example of (Monden 2012)

Figure 3.1 illustrates the kanban logic on a milk run with a pick up frequency of two times per day scheduled at 8:00 h and 20:00 h. The time for loading and unloading is omitted to simplify the figure. It is assumed that the reproduction of the parts triggered by the kanban arriving at 20:00 h of the last night is possible until the next vehicle arrives. To clarify the role of kanban in this example we show physical kanban cards, even if nowadays often eKanban cards are utilised in practice.

In contrast to the example, usually, more than one supplier is served on one tour in order to keep transportation cost low and realize at the same time high frequencies. Hence, the definition of a *round-tour mixed loading* system, a *milk run* or a *milk round* can be given as frequent "pick ups and deliveries at fixed times along fixed routes" (Baudin 2004). If it is necessary to exchange specialised containers, the tours usually start and end at the plant in order to ensure circulation of empty boxes which corresponds exactly to the physical kanban card circulation illustrated in Figure 3.1. However, if no containers are exchanged, and eKanban are used, the tour does not have to be a round trip.

Due to uncertain driving times, usually time slots instead of fixed times for the pick ups are arranged. For nearby part manufacturers these time windows are, for example, half an hour (cmp. (Bicheno and Holweg 2009)).

For the inbound use case there exist different types of milk runs depending on the distance between supplier sites and the customer plant. The easiest case are local milk runs, where the suppliers are close to the plant so that a milk run can be performed once or even several times per shift or day. This type exactly corresponds to the transport concept direct milk run (7) in Chapter 2.

If the suppliers are geographically clustered but far away from the plant, milk runs might still be possible: Either the truck does both, the line haul between cluster and plant and the milk run within the supplier cluster. Or the milk runs are performed through remote cross docks or consolidation centres, which build the interface to the long haul transportation with distinct vehicles (cmp. (Baudin 2004)). The first type corresponds to the direct milk run concept (7), the latter corresponds to the pre-/main-leg milk run (8) introduced in Chapter 2.

Supplier milk run parts are ideally supplied directly into a so called supermarket without any intermediate storage. A supermarket is an inventory store, close to production lines, necessary if a continuous flow – or stated differently, a one piece flow – is not possible (Rother and Shook 1999). Delivering directly into a supermarket, obviously, saves handling and storage capacity and makes inventories visible. But since the area at a production site is restricted, the capacity per product or product group is restricted and small supply volumes are necessary. If the vendors are located far away from the receiving plant, the parts are usually delivered into intermediate stores since the lots are too big for the supermarket, and the risk for delays due to the long driving distances is high. Another common system is a delivery in a so called collection centre close to the plant. In both cases, a kanban system similar to the described system can be established (for details see (Monden 2012)).

In general, the milk run concept along with a supplier kanban control is the result of the application of the described lean principles: It allows for a smooth flow due to small lots pulled by kanban signals and frequent supplies within constant cycles. As already mentioned in Chapter 2, it is very often applied within plants to feed the production lines with parts, and it can also be used for outbound logistics to supply distribution centres or regular customers (see (Baudin 2004)).

3.4 Operation Planning for Supplier Kanban Systems

A kanban system is a closed, decentralised self regulating system. That means, as soon as the number of kanban is determined, the system runs independently. Since the number of kanban restricts the WIP, the determination of the number of kanban and the adaptation over time is crucial (cmp. for example (Kotani 2007) and (Ohno et al. 1995)). For a detailed discussion of these methods we refer to Chapter 5. However, as a basis we first introduce the so called Toyota Formula that is often used in practice and was published for example in the first edition of (Monden 2012) in 1983.

A constant cycle withdrawal system, and hence a supplier kanban system, is defined by Monden (2012) through the order cycle expressed as the integer numbers $(a; b; c)$: During a days b deliveries are made, and each delivery will be delayed by c delivery cycles with a delivery cycle being defined as a/b. Correspondingly, in the example from Figure 3.1 the delivery cycle is expressed by $(1; 2; 1)$ because there are $b = 2$ deliveries during $a = 1$ days and the reproduction of the parts can be realized in the same cycle in which the supplier kanban arrive ($c = 1$). In the following we refer to a as *system cycle time* and to b as *frequency*.

Using these parameters, the *cycle time* K is given as $K = a/b$. The *lead time* L of the part consists of the time span during which the supplier receives the kanban, issues the corresponding production order, completes it and delivers the part to the plant. If the supplier holds finished goods on stock, the lead time is reduced and corresponds to the transport time. If an eKanban system is used, the lead time is not necessarily a multiple of a/b and consequently c is not an integer. In this case, c is given by $c = L/K$, otherwise it is $c = \lceil L/K \rceil$. The so called kanban lead time is $K + L = a/b + a/b \cdot c$.

Assuming an average part demand of d, the number of kanban N can be calculated by the so called Toyota formula:

$$N = \left\lceil \frac{d \cdot (K+L)}{M} + \frac{S}{M} \right\rceil = \left\lceil \frac{a \cdot (c+1)}{b \cdot M} + \frac{S}{M} \right\rceil \tag{3.1}$$

The parameter S corresponds to the safety stock, which is necessary due to demand and process uncertainties, while M stands for the container size. In practice, often an initial safety coefficient, which is iteratively reduced over time until supply breaks, is chosen to determine the safety stock.

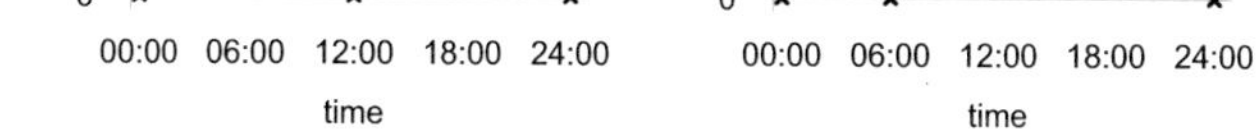

Figure 3.2: Inventory levels for a constant cycle time on the left chart and for varying cycle times on the right chart assuming a constant demand of one part per hour

There are two key conditions for reaching a low WIP level and to avoid the necessity of great safety stocks when determining the order and delivery times: The lead time must be reduced as much as possible, and to "the greatest possible extent the time intervals between one delivery to the factory and the next delivery must be constant" (Kotani 2007).

The example in Figure 3.2 illustrates that a constant cycle ensures not only levelled shipment sizes but also a lower WIP level since it is always the greatest time in between two deliveries, which determines the minimum number of kanban. A more detailed description of inventory effects in kanban systems follows in Chapter 7.

3.5 Milk Run Schedules and Heijunka Levelling

A sensible cycle length depends on the demand rate and the physical dimensions of the considered parts and, hence, is specific for every supplier plant relation. However, differing cycle times complicate the consolidation of more than one supplier on a milk run, which is essential for realizing high frequencies at acceptable transportation cost. For that reason, according to Monden (2012) the daily delivery frequencies at Toyota are restricted to multiples of 2 resulting in daily delivery frequencies of $\{2, 4, 6, 8, 10, 12\}$. Chuah and Yingling (2005) report from a Toyota Manufacturing plant in the US, that for a system cycle time of $a = 1$ day with a maximum delivery frequency of $b = 8$ only the small frequency set $\{1, 2, 3, 4, 8\}$ is allowed. That means all inter arrival times apart from the ones of frequency 3 are a multiple of 3 hours (see Figure 3.3).

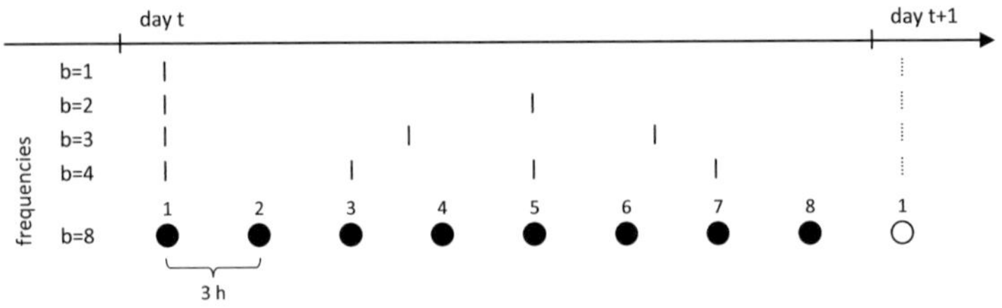

Figure 3.3: Pick up pattern with constant cycle times for the allowed frequencies of a Toyota Manufacturing plant in the US based on information of (Chuah and Yingling 2005)

The reduced number of frequencies increases the probability that two geographically close suppliers have the same frequency and can be joined to a milk run. A policy combining only suppliers with common frequencies is called Common Frequency Routing (CFR) (see (Chuah and Yingling 2005)). If also suppliers with different frequencies are allowed to be served on the same tour, this policy is called General Frequency Routing (GFR) (introduced in (Ohlmann et al. 2008)): For example suppliers with a frequency of 2 and 4 can be combined on every second milk round resulting in constant cycle times for both. The GFR policy is more complex from an operational point of view. The tours differ over the course of the day and, even in case of perfectly levelled inter arrival times at the supplier, the inter arrival times at the plant might vary slightly. The reason is that in two tours the remaining stops on the way to the plant might be different. On the other hand, GFR leaves a greater potential for consolidation and results in lower transport cost (a more detailed introduction follows in Chapter 5).

Chuah and Yingling (2005) and Ohlmann et al. (2008) give another reason, why perfectly equal inter arrival times at the supplier and the plant site can never be guaranteed in practice: differing operating hours at suppliers and the plant. In a kanban controlled system the, in some cases, considerable differences in inter-arrival times can lead to large deviations in shipment volumes. This results in higher inventories (remember Figure 3.2) and it clashes with the lean principle of a levelled use of the resources because the number of necessary vehicles and the amount of work in cross docks or shipping areas strongly varies over the course of the day. Hence, in both approaches it is proposed to split a supplier's daily demand into loads of equal size, while spreading the visits as equally as possible over the supplier's operating hours.

Neither Chuah and Yingling (2005) nor Ohlmann et al. (2008) report how this splitting into equal loads – also called transport lots – is realized within a kanban system. Therefore, we describe the application of a so called heijunka board at the supplier site in the following.

This concept is originally "a simple method for lot-sizing and production scheduling" (Furmans and Veit 2013) using kanban cards: It uses a levelling or heijunka board, which caps the maximal shipping volume and delays the shipment of supplier kanban parts exceeding the maximum lot to the next shipment. Figure 3.4 illustrates the mechanism: Let us assume that the daily demand is constant and amounts to 15 parts, with one part corresponding to one kanban, and three pick ups are scheduled at the beginning of period 6, 13 and 15. This schedule yields differing inter arrival times. The figure shows the situation at the beginning of period 13. Parts 6 to 10 are loaded and shipped to the receiving plant, while the parts 11 and 12, whose withdrawal was already signalised, are delayed to the next shipment because they exceed the shipping lot of 5 parts. If there are less than five kanban scheduled for a shipment – by putting the card into the corresponding slot of the heijunka board – the resulting transport lot is below 5.

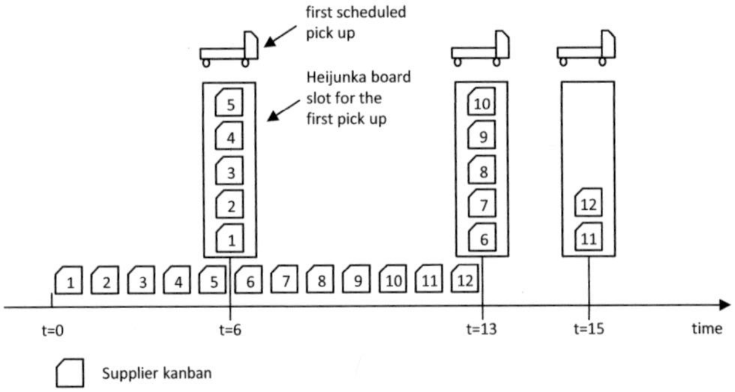

Figure 3.4: Illustration of the heijunka levelling mechanism applied to a supplier kanban system

This mechanism takes over two tasks: It ensures the levelling necessary due to varying – but planned – inter arrival times as shown in the example. It also caps the volumes in case of unplanned delays of the vehicle and demand variation. In a heijunka levelled system the transport shipment is guaranteed to never exceed the maximal shipping lot. In contrast to other levelling methods described in literature as the post office system and the smoothing of kanban collections, which was developed for physical kanban (see (Monden 2012)), this mechanism is established at the supplier site, not at the plant site. It can be easily realized without physical boards by using eKanban.

3.6 Milk Runs with a Cycle Time $a > 1$

In general, typical use cases for kanban controlled milk runs have a milk run cycle time – the time until a milk run plan is repeated for the first time – of $a = 1$ day. If the frequency b is greater than 1, usually only even numbers are allowed (see (Kotani 2007) and (Monden 2012)). Cases with $a > 1$ are considered according to Kotani (2007) as very special.

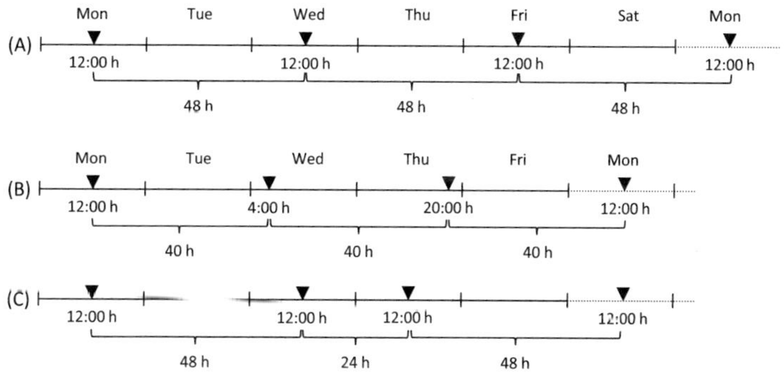

Figure 3.5: Pick up times for different definitions of schedule consistency: (A) arrival time and inter arrival time consistent schedule (B) inter arrival time consistent schedule and (C) arrival time consistent schedule

However, in Chapter 2 we showed that there exist many milk run systems in which the demand volumes only allow cost efficient frequencies smaller than once per day, for example, 3 times per week. Figure 3.5 (A) shows that in case of a six day working week, schedules can be constructed with a constant cycle time and with consistent arrival times. However, in case of a five day working week the constant cycle time requirement leads to strongly deviating pick up and delivery times (B). This often clashes with the opening hours of suppliers or the consolidation centres of the logistics provider, it might be in conflict with the ban on night time driving, and the resulting schedules are difficult to memorise.

Hence, in these cases, a deviation from the constant cycle time leading to arrival time consistent schedules, as shown in example (C), might be beneficial. In this case, the heijunka levelling mechanism described in the preceding section can be applied to reach volume consistency. However, to the best of our knowledge there exists no work considering kanban controlled milk runs with system cycle times longer than a day.

For these cases, apart from milk runs, also groupage services with delivery profiles described in Chapter 2 might be applied, if a supplier kanban cycle is in place. The run time of groupage services usually is longer and hence, more kanban are needed. However, in case of suppliers geographically scattered and far away from the receiving plant, groupage services might lead to lower operational cost than a milk run service.

3.7 Summary

To understand the logic of milk runs within lean manufacturing one needs to be aware of the key lean principles and the kanban concept: A low WIP is a key goal in lean manufacturing and small lot sizes sent in constant cycles lead to a low WIP and a smooth flow of parts. However, in supplier kanban systems cost efficiency for the routing part forbids a one piece flow and, for example, differing opening hours, do not allow for perfectly constant inter arrival times of the milk run.

In case of varying inter arrival times, a levelling mechanism is applied to reach consistent shipping volumes (see i.e. (Ohlmann et al. 2008) or (Chuah and Yingling 2005)). We described how such a mechanism can be realized using a heijunka board at the supplier site.

The very same mechanism can also be used to level the shipping volumes for weekly milk runs with consistent arrival times and – purposefully – varying inter arrival times in order to be in line with opening hours and other restrictions such as night drive bans. To the best of our knowledge, in literature there exists no work addressing the problems arising in kanban supplier systems with shipping frequencies below once per day, although if the latter is a very common situation in industrial practice.

In order to be able to tactically plan milk runs, one balances cost for WIP and cost for transport considering a "pure" kanban mechanism or a kanban mechanism with heijunka levelling. Hence, in this thesis we consider the question of how this tactical milk run scheduling decision can be modelled focussing on the case of cycle times bigger than one day, which is barely considered in lean literature.

4 Milk Run Design

In Chapter 2 and Chapter 3 we showed that regularity and especially regular tours are beneficial within and outside of the lean world for a couple of reasons. On the other hand, according to our experience groupage services without any regular components are still the most common transport concept used in industrial practice. In case of Bosch Homburg – our partner of the case study introduced in Chapter 10 – groupage services executed by different area contract forwarders are the only standard road transport concept: The main reason is the simple dispatching for the material planners and the ease to control cost in case of varying volumes.

In order to understand the complexity of designing milk runs for regular suppliers on a tactical level, we describe the problem that practitioners face when they execute this task manually: We follow a common approach decomposing the decision in three easier subproblems, which are typically solved sequentially using heuristics for every step: (1) the transport concept assignment, (2) the frequency assignment, and (3) the milk run scheduling decision. In Section 4.4 we propose an integrated version of the Milk Run Design (MRD) decision including subproblem (1), (2) and (3). It is the central definition of this thesis and serves as starting point for the following chapters dedicated to the question how the decision can be modelled adequately. As usual, we conclude the chapter with a short summary.

4.1 Transport Concept Assignment

In the first step, every regular supplier is assigned to a transport concept. It is determined if the supplier is served by a milk run with one or more suppliers, by a point-to-point transport with a fixed cycle – not necessarily in line with the milk run cycle time – or a fixed schedule, or if the supplier is better served by groupage service through an area forwarding network.

The assignment depends for very expensive or critical parts – i.e. parts delivered according to the just in time or just in sequence concept –

on thoughts concerning reliability: For these parts, the higher reliability resulting from milk run shipments might be desirable irrespective of the operational logistics cost. For all other parts, the decision is assumed to be driven by operational cost.

As described in Chapter 2, the cost for a shipment through the area forwarding network and by point-to-point transports can be determined independently for every supplier plant relation by looking up the tariff in a tariff table. The cost for a milk run depends on the distance and the duration of the whole tour. Hence, the cost for one supplier depends on the decision if, for example, a neighbouring supplier is also allocated to the milk run concept and is served on the same tour.

That means, the transport concept decision and the definition of the milk runs described in subproblem (3) are ideally taken at the same time in order to be able to balance the cost for the three concepts. Otherwise, the transport concept assignment depends on good milk run cost estimations or must be taken heuristically by simple rules considering the shipment structure and geographical position. The latter is proposed in the recommendation of the VDA (Projektgruppe Standardbelieferungsformen 2008) introduced in Chapter 2 and is presented as a common approach used by practitioners in (Branke et al. 2007). A detailed description of these rule based decisions follows in Chapter 5 and Chapter 9 respectively.

4.2 Frequency Assignment

The frequency assignment refers to the step of determining the number of shipments between a supplier and the receiving plant per day or week. A high frequency results in small transport lots and, hence, in low inventories and inventory related cost. On the other hand, frequent shipments lead to high transport cost. Both cost types, inventory and transport cost need to be balanced in order to be able to define a cost optimal frequency for every supplier.

Since the chosen kanban control and reorder strategy determine the shipment volumes for a given frequency or delivery pattern, the frequency assignment implicitly determines the shipment volumes. The latter could be considered as an input parameter for the transport concept decision. On the other hand, the transport cost for a supplier depends on the assigned transport concept and in case of milk runs even on the concrete milk run schedule. Consequently, a cost optimal frequency decision is ideally taken at the same time as the transport concept and milk run scheduling decision.

Practitioners often determine the frequency before assigning a transport concept. This can be explained by the fact that the decision is often influenced by limiting resources: These resources require an upper limit for the maximum lot size which results in a minimum frequency. An example of limiting resources can be the number of available specialized container or loading aids and the reserved areas in the warehouses or supermarkets of the supplier or the receiving plant.

Furthermore, some positive effects of high frequencies – such as a smooth flow of goods, a good stock visibility or smaller necessary storage areas – are difficult to evaluate in monetary units. Hence, the definition of a minimum frequency is not necessarily only based on operational cost.

4.3 Milk Run Scheduling

Even if we argued above that the three decisions should be ideally taken simultaneously, we assume in this section that we determine milk run schedules separately. That means, we consider only the suppliers assigned to the milk run concept in step (1) and assume that the frequency decision (2) is already taken. Hence, a fixed pick up cycle – i.e. every three hours or every second day – or a set of feasible pick up patterns – i.e. Mon-Wed-Fri, Mon-Wed-Thu, etc. – is given for all of the "milk run suppliers". Please note, that since a cycle can be converted into a pattern, we do not always explicitly refer to fixed cycles in the following.

The following decisions are made in the actual milk run scheduling step: (1) every supplier is assigned to a feasible pattern, i.e. every visit is assigned to a period or day, (2) every visit is assigned to a vehicle available within the corresponding period, (3) the sequence of visits for every vehicle is defined and (4) the start of service in the depot and for every visit is fixed.

These decisions are taken considering the number of available vehicles per period (if restricted), the capacity of the respective vehicles, the maximum allowed driving time of a driver and the opening hours of the stops. During the scheduling decision described in step (4) further requirements considering arrival time or inter arrival time consistency resulting from a kanban control strategy or other time consistency requirements are considered. The goal is to minimize the transport cost by minimizing the travel distance (or travel time) and the number of necessary vehicles. There might exist further requirements described in Chapter 2, such as driver consistency, or a slightly different minimization goal.

Suppliers assigned to groupage service or to point-to-point transports can be treated independently when assigning the pick ups to periods. The planner can either assign fixed delivery profiles or fixed cycles. However, he/she can also waive regularity requirements and leave the concrete scheduling to the daily material planning.

4.4 Milk Run Design Decision

The three preceding sections described the decomposed decision process for the Milk Run Design (MRD) decision common in practice. However, due to the discussed interrelationships between the three subproblems, they are considered ideally in an integrated fashion.

In the following we define the Milk Run Design problem by specifying the scope of the decision, by characterizing typical input parameters and by discussing the length of planning intervals.

4.4.1 Scope of Decision

Summarizing subproblem (1) to (3) in an integrated approach addresses the following decisions simultaneously:

1. Assign every supplier to one of the transport concepts, milk run, point-to-point transport or groupage service.

2. Assign a frequency and determine a feasible pattern and the corresponding volumes in accordance with the kanban or reorder policy.
3. Assign all milk run suppliers to a feasible pattern and, hence, every stop to a period.
4. Assign every visit to a vehicle with sufficient capacity available at the corresponding period.
5. Determine a sequence of visits for every vehicle at every period.
6. Fix the start of service of every vehicle and the start of service at every stop.

The decisions are taken while minimizing the transport cost and the cost for inventory holding at the plant site and – if possible, for example, in case of a supplier kanban system in place – at the supplier site. The transport cost incorporates the cost for groupage service, milk runs and point-to-point shipments. For the milk run part, capacity and time window constraints need to be considered.

4.4.2 Overview of Typical Input Data

The typical input data for the Milk Run Design problem for the inbound use case includes demand, inventory and transport related cost, and information about sensible base periods or system cycle times. In the following they are listed and shortly discussed.

The part demand of the plant for every supplier is typically given as a consumption rate per period. It is expressed in shipping units – for example, pallets per day or standard boxes per hour or shift. Furthermore, the cost for holding parts on stock are given, for example, as Euro per unit time on stock (see also Chapter 7).

As described in Chapter 2, the transport tariffs for groupage services are usually agreed between a consignee and an area contract freight forwarder and can be looked up in the tariff table.

In order to be able to plan milk run tours we need fleet, vehicle and driver characteristics, such as number of available vehicles (if restricted), vehicle capacities, driving times and distances within the inbound network, and the corresponding cost parameters per driving time and distance unit as well as a fixed cost factor for a vehicle if applicable.

As already mentioned in Chapter 2 and Chapter 3, the system cycle time is the time or the number of base periods – in the following base period is also referred to as period – elapsed until every tour is serviced at least once or until the delivery profiles are repeated. If milk runs are used to feed the production, the system cycle time can be seen as a result of the decision process. However, in the inbound use case the system cycle time is usually given by the maximum acceptable time between two shipments for the suppliers considered as regular. Literature shows that in most of the cases one or two days or one working week is sensible.

If the system cycle time is one day, the base period length results from dividing the system cycle time by the maximum delivery frequency. At Toyota for example according to Monden (2012) the maximum delivery frequency for supplier kanban systems is 12 and hence, the base period is 2 hours. If the system cycle time is one week, the natural base period is one day. However, it might make sense to set the base period to half a day or to the length of a shift if some suppliers should be serviced twice a day or once per shift. If the system cycle time is longer than a week, the advantages of the regularity get more and more lost.

For this thesis we assume that a sensible system cycle time and the corresponding base period are derived from the circumstances, such as demand rates per day and driving times, and are already given when the Milk Run Design decision is taken.

4.4.3 Planning Intervals

Milk runs and delivery profiles should be valid for a certain period in order to achieve the positive effects – like better tariffs or coordination and learning effects – described in the preceding chapters. Hence, the Milk Run Design is a tactical planning task producing schedules valid for several weeks up to several months.

The planning intervals can be either fixed to the same length that the validity period has, or the planning can be executed whenever input parameters change. Furthermore, the planning might be executed for "what if" analysis in order to answer questions such as: What is an appropriate dimension of a supermarket or storage area? Or what is an appropriate number of specialized containers?

4.4.4 Advantages of Automated Planning

Due to the complexity of an integrated decision, an IT based decision support system seems beneficial: It can be expected that the results of the Milk Run Design can be produced quicker compared to a manual sequential approach and that an integrated approach yields better results in terms of operational cost and in terms of desirable regularities. Furthermore, the results are considerably less dependent on the qualification and the experience of the responsible planner.

Furthermore, an IT based decision support system is a prerequisite for frequent reviews of the current plans considering input parameters, which slightly change over time. The planning approach is necessary to identify the moment, when an adapted milk run plan is better than the current one. Obviously, if an input parameter changes significantly, a new plan can be put in place immediately. It is also possible to test different order policies or kanban control scenarios within a simulation framework. This is important because a good automated order policy or a properly dimensioned kanban control system resulting in regular demands is crucial for successfully applying regular transport schedules.

4.5 Summary

We can summarize the goal of the Milk Run Design as the definition of cost efficient, regularly recurring transport schedules for suppliers with a regular demand incorporating:

- a regular supply via the area forwarding network by determining recurring delivery profiles,
- regular Milk Run Schedules fulfilling certain consistency requirements within the milk run cycle time for the supply with milk runs,
- and regular full truck load shipments.

Due to many influencing factors and the interdependencies between the underlying subproblems, it is difficult to define such a schedule manually. An IT based decision support system is necessary to manage the additional complexity for the consignee compared to the case of using only groupage services and point-to-point transports.

5 Related Optimization Approaches

The literature related to the introduced Milk Run Design problem is extremely diverse: It ranges from optimal lot sizing over network design to vehicle routing and ends with literature specifically addressing lean manufacturing. Historically, these areas have often been considered separately. However, the "increase in computational power and the development of efficient metaheuristics" (Schmid et al. 2013) offers the opportunity and strengthens the trend to consider logistics decisions in a more integrated way than before.

The integration of consecutive planning problems is done in all described research areas: Transport aspects have been integrated into inventory and lot sizing models, inventory aspects have been integrated into vehicle routing models, and transport and lot sizing decisions have been integrated into network design approaches. These developments and the fact that in practice concepts are not defined precisely – see also Chapter 2 – has led to an ambiguous use of terms and concepts in literature.

A good example are the differing terms used for what we defined as transport concepts in Chapter 2: Blumenfeld et al. (1985) refer to it as routes through the transport network or distribution strategy, Miemczyk and Holweg (2004) talk about an inbound scheme, Berman and Wang (2006) use the term distribution strategy, Branke et al. (2007) introduce the term transport channel, Dekker et al. (2012) also apply the term transport concept, while Rieksts and Ventura (2008), Kempkes et al. (2010), and Harks et al. (2014) use the word transport mode. However, the latter is classically used to describe if a transport is executed by plane, ship, truck, rail, barge, or pipelines (see (Meixel and Norbis 2008)).

In this chapter we classify the relevant theoretic literature and the approaches for specific use cases from different research areas with regard to the requirements of the Milk Run Design. In order to give an orientation with respect to the Milk Run Design problem, we introduce the related approaches following the structure of the subproblems: Transport concept assignment, frequency assignment, and milk run scheduling.

However, due to the reasons given above, the borders become blurred in some cases.

Since this thesis is dedicated to deriving models for the Milk Run Design and not necessarily to propose solution approaches applicable to big instances, we focus on models in this chapter. However, without introducing basics, we give short information on solution methods for readers who are familiar with common solution approaches of discrete optimization.

After a short summary at the end of this chapter, we assess the models and point out the research gaps related to the Milk Run Design in Chapter 6 and derive the resulting need for action upon which the succeeding chapters are based.

5.1 Transport Concept Assignment

The choice of the appropriate transport mode – namely transport by plane, ship, truck, rail, barge or pipelines – has been studied for many years, as well as the carrier selection for the corresponding mode. An extensive overview of the literature over the past two decades is given in (Meixel and Norbis 2008). In the following, we focus on literature addressing different concepts for transports by road.

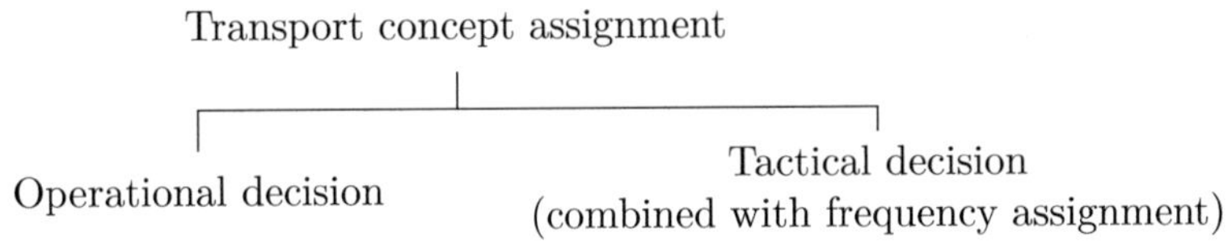

Figure 5.1: Top level classification scheme of literature related to the transport concept assignment

For reviewing the work on transport concept selection we differentiate between the short-term, operational decision, usually taken when the pick up and delivery date of an order is already fixed and the tactical decision resulting in a transport concept assignment for a supplier valid for a certain period (see Figure 5.1). The latter is usually combined with a frequency assignment decision. Hence, the corresponding literature is introduced in the following section.

In practice the operational decision on the best transport concept is taken based on simple rules considering the weight, volume or loading metres of an order. Usually, the alternatives are restricted to point-to-point transports in case of FTL shipments, groupage services for LTL shipments and CEP for small packages. The VDA gives the recommondation to consider shipments above 11 loading metres as FTL, below 11 loading metres as LTL, and below 31.5 kg as small packages. Following Branke et al. (2007) we refer to these assignment schemes as weight based allocation heuristics (WA).

In contrast to the simple case described in the preceding paragraph, Branke et al. (2007) introduce a more complex setup for the transport concept assignment in their work called transport channel selection: They consider – besides dedicated point-to-point transports and groupage services – dedicated direct tours and indirect tours via a hub. Please note that for the sake of consistency we use the terms introduced in Section 2.1.2 to describe the concept. The terms originally used in the articles can be found in brackets in column 3 of Table 5.1, which gives a short overview on the work considered in this section. The latter two concepts require ad hoc tour planning and, hence, the definition for the operational transport concept assignment task is "assigning shipments to transport channels, and optimizing the routes within each channel" (Branke et al. 2007).

The heuristic proposed to solve this problem is an evolution based algorithm for assigning the transport channels and a parallel savings heuristic with a 2-opt local search which is applied in the evaluation phase to the channels requiring routing decisions (a short introduction to vehicle routing problems follows in Section 5.3.1). The authors compare their approach to a WA heuristic, in which all customers with an order weight below w_{min} are served by groupage service, the ones with an order weight above w_{max} are served by a dedicated point-to-point transport, while the ones whose volume weight is in between are assigned to dedicated direct tours. The evaluation was performed on real-world data with 180 to 300 customers considering realistic tariff tables and vehicle costs. The evolutionary algorithm (EA) based heuristic decreases the transportation cost compared to the WA results on average by 8% and maximally by 12%.

Kempkes et al. (2010) propose a network based formulation to select for an order either point-to-point transports, groupage service, or CEP. In this case, the interdependence between the assignments of suppliers arises from the fact that different tariff tables for the pre-leg and the main-leg are considered for the groupage service. This means that the decision to assign an order to a direct point-to-point transport might result in smaller cost for this relation but higher cost in total since on a main-leg tour a low capacity usage rate results from the shift. The solution approach involves a construction and an improvement heuristic and the use of a commercial MIP solver injected with the starting solution generated by the heuristics.

The authors showed that savings of up to 30% could be realised for one out of seven areas of a case study considering the area forwarding network of an original parts manufacturer. Further cost reductions could be achieved by introducing delivery patterns on a tactical level. However, this part concerning frequency assignments is described in Section 5.2.

The model introduced in (Hosseini et al. 2014) is based on the model of (Hosseini et al. 2014) and it extends a network flow formulation by vehicle routing in order to simultaneously plan point-to-point transports, point-to-point transports over a hub and dedicated direct tours. The authors do not consider tariff based cost for shipping via a hub, but distance related cost for both links from supplier to the hub and from the hub to the plant – this corresponds in the classification of Chapter 2 to a special case of dedicated indirect tours with all pre-leg and main-leg tours consisting of just one source and destination. The hybrid approach of harmony search and simulated annealing was evaluated on artificial instances but not on real-world data.

All introduced approaches are for the use on a daily basis and consider only ad hoc concepts, that means, no concepts with regularities are considered. However, if no regular tours such as milk runs are in place and arranged with suppliers and freight forwarders, an assignment to this concept is not possible. Hence, the Milk Run Design problem introduced in the preceding chapter addresses the tactical counterpart of the operative transport channel decision as defined in (Branke et al. 2007).

Table 5.1: Overview of related literature part I: Operational transport concept assignment

Author	Considered transport cost	Considered transport concepts	Solution approach
Branke et al. (2007)	tariff tables for groupage service, distance based tariffs for all other concepts	(1) dedicated direct tour [milk run] (2) point-to-point transport [dedicated truck] (3) dedicated indirect tour [delivering over a hub] (4) groupage service [delivery service]	WA heuristic and simple "first routing second improving" heuristic as benchmark for their approach: evolution based heuristic for assigning the transport concept and a parallel savings heuristic with a 2-opt local search for the routing part
Kempkes et al. (2010) part I: transport mode	group of complex and simple tariff structures	(1) point-to-point transport [direct transportation] (2) groupage service [area forwarding] (3) courier, express and parcel service	multi period multi commodity network design formulation considering different tariff structures and discount schemes with symmetry breaking, solution procedures: construction heuristic followed by an improvement heuristic, MIP solver with starting solution from heuristics
Hosseini et al. (2014)	distance based cost for all links	(1) point-to-point transport [direct shipment] (2) point-to-point transports over a hub [shipment through cross-dock] (3) dedicated direct tour [milk run]	network flow model extended by vehicle routing for dedicated direct tour considering many suppliers and few plants, solution procedure: hybrid approach of harmony search and simulated annealing

Delivery profiles or resulting milk run plans would be the input for this operative planning step. An integration of this input in the network flow models is straight forward.

5.2 Frequency Assignment

When assigning a frequency to a supplier plant relation, the transport and inventory cost need to be balanced as described in Section 4.2. This corresponds to solving a replenishment problem or more generally an economic lot size model.

The inventory cost not only depends on the frequency but also on the order policy or kanban system in place because they determine the lot size for a given frequency. In the following we differentiate the literature overview by push based order policies and pull based kanban systems (see Figure 5.2).

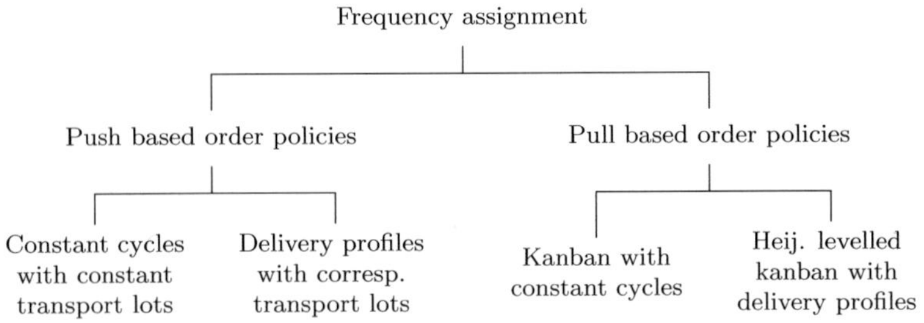

Figure 5.2: Top level classification scheme of literature related to the frequency assignment

As long as the transport cost for a relation only depends on the order for the relation itself – for example in case of point-to-point transports – the frequency can be assigned without solving routing problems. In this section we focus on literature which assumes that this is fulfilled or that the cost for tours are approximated. In Section 5.3 we introduce work explicitly solving vehicle routing problems.

5.2.1 Push Based Order Policies

In this section we introduce basic models for determining frequencies and lot sizes relevant to the Milk Run Design problem. The goal of the models in the first subsection is to determine schedules with constant cycles and constant transport lots, while the ones in the second subsection determine delivery profiles with varying inter arrival times and the corresponding lot sizes (see Figure 5.2).

Constant Cycles with Constant Transport Lots The classic economic lot size model was introduced by Harris in 1913 (for a reprint see (Harris 1990)), and nowadays it can be found in every text book on production and logistics.

For the model it is assumed that the point of consumption has a constant demand rate of d items per unit time, that the lead time is 0, the lot size q is the same for every order, the initial inventory is 0 and the planning horizon is infinite. The products and time are assumed to be continuously divisible. A fixed setup cost c^o – in our case including transport cost – is incurred for every order and a linear inventory holding cost c^h for every unit on stock per unit time is accrued (for an introduction see for example (Simchi-Levi et al. 2014a) or (Hopp and Spearman 2001b)).

Given these assumptions, the economic order quantity (EOQ) or economic lot size minimizing the overall transport and inventory cost are given by:

$$(EOQ) \qquad q^* = \sqrt{\frac{2 \cdot c^o \cdot d}{c^h}} \qquad (5.1)$$

The inventory level over time follows the perfectly regular sawtooth pattern showed in Figure 5.3 with a maximum inventory of q^* and a cycle time – in case of supply from external sources also called order interval – of length q^*/d. Hence, the zero-inventory-ordering property, saying that a shipment must arrive precisely at the moment when the inventory drops to zero, is fulfilled. Under the assumptions described above, this property is valid for every optimal order policy.

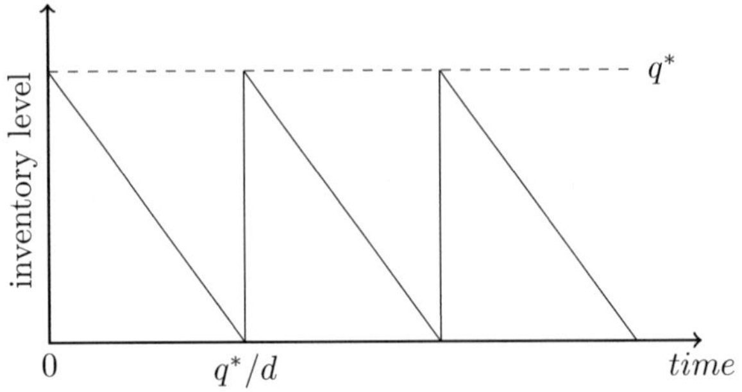

Figure 5.3: Inventory level as a function of time for an EOQ solution

The sensitivity analysis of this model shows that holding and setup costs are rather insensitive to lot sizes or order intervals. This motivates the so called power-of-two policy restricting the order intervals t to be a power-of-two multiple of a fixed base period t^b

$$t = t^b \cdot 2^k \qquad k \in \{1, 2, 3, \cdots\} \tag{5.2}$$

Such a policy is applied to allow for an easier consolidation of orders to achieve a better truck usage. One can show that the increase in inventory and setup cost for an optimal power-of-two interval is guaranteed to be no more than 6% above the cost for the optimal order interval (see (Simchi-Levi et al. 2014a) or (Hopp and Spearman 2001b)). If this is compensated by lower transportation cost, the restriction of the interval length is worthwhile. This approach is very similar to the one described by Chuah and Yingling (2005) and Monden (2012) (see Section 3.5), who report that the frequencies are restricted to enable consolidations when planning milk runs.

If the parts are only needed for a limited period and the planning horizon is not infinite but of length T, an optimal order policy is very similar to the one of the infinite case: One places f^* orders of equal size at equally spaced points in time. The optimal number of orders f^* is $\lfloor \alpha \rfloor$ or $\lceil \alpha \rceil$ depending on which value yields smaller cost with

$$\alpha = T \cdot \sqrt{\frac{c^h \cdot d}{2 \cdot c^o}} \tag{5.3}$$

The time between two orders can be given by T/f^*, and the lot size – following the zero-inventory-property – corresponds to $(T/f^*) \cdot d$ (see (Simchi-Levi et al. 2014b)).

If the setup cost c^o is shared amongst multiple items, in literature the problem is classified as joint replenishment problem (JRP). This is for example the case if more than one item is ordered from the same supplier, or if a volume discount for the main leg transport is granted. Whenever the demand is no longer static but varies to a known extent over the planning horizon, the problems are classified as deterministic dynamic lot sizing problems (LSP). If parameters such as demand or lead times are not known, one talks about stochastic LSP.

Due to the vast literature in this research area we focus in this section on a selection of relevant approaches listed in Table 5.2, which all explicitly consider transport cost. For further reading we refer to (Khouja and Goyal 2008) or (Glock 2012) considering joint replenishment lot size problems, to (Robinson et al. 2009) reviewing coordinated deterministic LSP models, and to (Tempelmeier 2007) being concerned with the stochastic variant.

To the best of our knowledge Blumenfeld et al. (1985) were the first who extended the EOQ formula to the case of point-to-point transports being charged by a fixed value c^o: The optimal shipping quantity is the minimum of the EOQ formula and the vehicle capacity. Speranza and Ukovich (1994) extend this setup by considering the replenishment of different items from the same supplier. The authors furthermore restrict the shipments to a certain time of the day so that the inter arrival times between shipments are always multiples of a day – or stated differently – they assume that time is discrete. In this work different ways of assigning frequencies to items and consolidating items to shipments are compared by solving the corresponding integer linear programs.

Table 5.2: Overview of related literature part II: Frequency assignment with a constant cycle

Author	Considered cost	Considered transport concepts	Decision on frequency	Solution approach
Blumenfeld et al. (1985) B, Blumenfeld et al. (1987), Burns et al. (1985)	inventory related cost and transport cost	(1) point-to-point transport [direct shipping] (2) indirect: point-to-point transport to a cross dock and point-to-point transport to the destination [shipping via warehouse] (3) master tours or dedicated direct tour [peddling]	shipment quantity / frequency per transport concept	for every transport concept an EOQ based formula to determine shipment frequencies is developed, the formulas are based on approximations (peddling regions) and simplifications (inventory holding cost), based upon these formulas an approach to select the best transport concept for small instances is proposed
Speranza and Ukovich (1994)	inventory related cost and transport cost	point-to-point transport [single link]	optimal frequency per item with shipments restricted to certain times of the day – resulting in inter arrival times being multiples of a day	Integer and mixed integer linear programming models for different consolidation strategies: consolidation only for items with common frequencies or consolidation of shipments assigned to the same period
Swenseth and Godfrey (2002)	inventory related cost and approximation functions for tariff tables	(1) point-to-point transport [FTL] (2) groupage service [LTL]	optimal order quantity / frequency	approximated functions for estimating transport tariffs incorporated into inventory replenishment models

continued

Table 5.2: Overview of related literature part II: Frequency assignment with a constant cycle (continued)

Author	Considered cost	Considered transport concepts	Decision on frequency	Solution approach
Berman and Wang (2006)	inventory related cost and transport cost based on distance between two nodes	(1) point-to-point transport [direct] (2) master tours or dedicated direct tour [milk run (peddling)] (3) indirect: point-to-point transport to a cross dock and point-to-point transport to the destination [cross dock]	optimal order quantity / frequency	nonlinear integer programming model solved by greedy heuristic to generate initial solution and a Lagrangian relaxation heuristic and branch-and-bound algorithm, dedicated direct tours are considered as peddling regions determined by visual observation, in which every region is treated as a single supplier
Rieksts and Ventura (2008)	inventory related cost and tariff tables for groupage services	(1) point-to-point transport [TL] (2) groupage service [LTL] (3) mixture of both for the same relation	optimal order quantity / frequency	exact algorithms to determine optimal order policies for both concepts and for a mixture of both concepts

Swenseth and Godfrey (2002) incorporated approximation functions for transport tariffs into inventory replenishment models. Later, Rieksts and Ventura (2008) proposed an exact algorithm to determine the optimal order quantity and frequency for groupage service with tariff tables and for a mixture of groupage service and point-to-point transports. There exist further extensions, such as models considering also quantity discounts granted by the suppliers (see for example (Konur and Toptal 2012) or (Birbil et al. 2014)) or considering a two stage lot sizing problem (see for example (Jaruphongsa et al. 2007)). However they are beyond the scope of this work.

The exact EOQ formula for point-to-point transports introduced above was developed as one part of a decision support system for General Motors (GM) described in (Blumenfeld et al. 1985), (Burns et al. 1985), and (Blumenfeld et al. 1987): The goal of the application was to determine a so called distribution strategy, that is a shipment size / frequency and a route through the network, for every relation. As distribution strategies – corresponding to our term of transport concept – they considered point-to-point transports, master tours, and shipments via a transshipment point (comparable to groupage services with distance related cost on all links). Instead of modelling the problem as mathematical program, the authors proposed simplified formulas which approximate for example the master tours as so called peddling regions with a rough estimation for the local routing part or which simplify the inventory cost. These easy to use formulas supported the decision makers at GM to analyse the trade-off between inventory and transport cost for different concepts so that they could realise consolidation effects for the logistics network.

Later, Berman and Wang (2006) proposed a nonlinear mixed integer model minimizing inventory and transport cost considering the same transport concepts. Instead of solving a routing problem, the master tours were also approximated as peddling regions determined by visual observation and treated as a sinlge supplier. The authors proposed a greedy heuristic to generate an initial solution and a Lagrangian relaxation heuristic and branch-and-bound algorithm to solve this nonlinear problem.

Delivery Profiles with Corresponding Transport Lots The classic EOQ formula (5.1) starts from the premise that the products are needed for a long time and that products and time are continuously divisible. In contrast, Kovalev and Ng (2008) introduce a discrete variant of the EOQ model with a finite horizon: In such a setup, constant order intervals only result if the optimal number of order placements – called frequency f – is a divisor of the planning horizon T. For the other case, the authors showed that a minimum inventory level can be reached for the optimal frequency f^* if one places only orders either of size $d \cdot \lceil T/f^* \rceil$ or orders of size $d \cdot \lfloor T/f^* \rfloor$. Since the zero-inventory-property must apply, the inter arrival times also take at most two different values. The sequence in which the orders are placed has no impact on the cost. Based on the solution method proposed in (Kovalev and Ng 2008), Li et al. (2007) suggested an improved variant to determine the optimal frequency f^*.

In the work of Kovalev and Ng (2008), the proposed discrete EOQ model is not aimed for generating cyclically repeated delivery profiles. However, the optimal order schedule can be applied cyclically assuming that the planning horizon T corresponds to the system cycle time – for example one week.

Fleischmann (1999) also aims for the generation of cyclic delivery pattern with possibly varying inter arrival times. However, in this setup the length of the cycle time is the result of the optimization process. The shipping amount is determined in a way that the inventory and transport cost are minimal while the capacity of the vehicle executing the point-to-point transport is respected. The author derived optimality properties leading to inventory minimal pattern.

Besides selecting the cost optimal transport concept on the operational level (see Section 5.1), Kempkes et al. (2010) proposed to introduce delivery profiles for area forwarding and point-to-point transports minimizing inventory and transport cost. The same setup was investigated in Schöneberg et al. (2010) and Schöneberg et al. (2013) (see also Table 5.3).

In this problem setting, the costs for area forwarding are represented by two different tariff tables for the pre-leg and the main-leg (see also Section 2.3). The cost advantages are caused by spatial and time-based consolidation effects on the pre-leg and the main-leg and by shifts to point-to-point transports due to time-based consolidation at the suppliers.

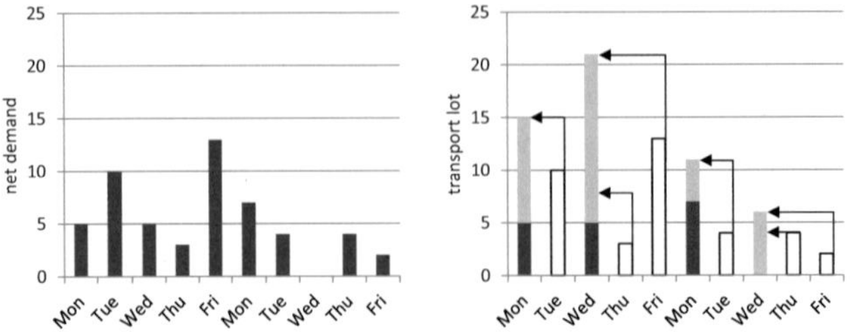

Figure 5.4: Generation of a delivery schedule by applying a delivery profile with delivery days Monday and Wednesday to a net demand forecast (Schöneberg et al. 2013)

Figure 5.4 shows the operational task of generating a delivery schedule by applying a delivery profile to net demands: On the left hand chart, net demands are plotted over their respective due date. On the right hand chart, all net demands until the next scheduled shipment are consolidated to a transport lot. This leads to a better capacity utilization on the pre-leg or allows a shift to point-to-point transports and, at the same time, levels out the order volumes. The selection of good delivery profiles then leads to additional consolidation effects on the main-leg.

Kempkes et al. (2010) included the delivery profiles in their network based model for the operational decision by forbidding links not in line with the available profiles (more details can be found in (Kempkes 2009)). In order to be able to solve bigger instances, (Schöneberg et al. 2010) proposed a model focused on the tactical task to assign delivery profiles: The authors also used a multi period multi commodity flow network model. To make the model computationally tractable a decomposition scheme along with preprocessing techniques and primal heuristics is proposed.

The approach of (Kempkes et al. 2010) was evaluated on the same instances as the operational transport concept assignment. The assignment of delivery profiles led in all seven areas to significant cost savings, while the combination of delivery profiles and the transport concept assignment yielded cost savings of up to 59% with an average of 34%.

Table 5.3: Overview of related literature part III: Frequency assignment with delivery profiles

Author	Considered cost	Considered transport concepts	Decision on frequency	Solution approach
Fleischmann (1999)	inventory related cost and transport cost	point-to-point transport [single link]	optimal delivery profile determined by cycle length and frequency per cycle with varying (discrete) inter arrival times for multiple products with free choice of the cycle length	optimality of dominant patterns is analytically proven
Kovalev and Ng (2008)	inventory holding cost and fixed setup cost	transport concepts with fixed transport cost	optimal frequency and lot sizes for finite horizon assuming discrete time delivery profiles	search based on the proof of convexity of the objective function
Kempkes et al. (2010) Part II: Delivery Profile	inventory holding cost and complex tariff structures with discount schemes for transport as in groupage service	(1) point-to-point transport [direct transportation] (2) groupage service [area forwarding]	delivery profile	same approach as described under (Kempkes et al. 2010) Part I in Table 5.1 considering delivery profiles assigned to every supplier

continued

Table 5.3: Overview of related literature part III: Frequency assignment with delivery profiles (continued)

Author	Considered cost	Considered transport concepts	Decision on frequency	Solution approach
Schöneberg et al. (2010)	complex tariff structures and inventory holding cost	(1) groupage service with different tariff tables on links of pre-leg and main-leg [area forwarding] (2) point-to-point transports [fulload]	delivery profile with dispatch rule to aggregate demand until the next delivery	multi commodity network flow formulation, as solution approach a decomposition scheme, preprocessing techniques and primal heuristics and the use of a MIP solver are proposed
Schöneberg et al. (2013)	inventory holding cost and transport cost given in tariff tables	(1) groupage service with different tariff tables on links of pre-leg and main-leg [area forwarding] (2) point-to-point transports [fulload]	delivery profile with dispatch rule to aggregate demand until the next delivery	simplified formulation of (Schöneberg et al. 2010) extended to a two stage stochastic model, solution framework consists of scenario reduction techniques, a decomposition technique, a genetic algorithm and a MIP solver
Meyer et al. (2011)	transport cost given in tariff tables and inventory holding cost	groupage service network with one tariff table for pre and main-leg	delivery profile with dynamic lot sizing model levelling out demands over the course of the week	dynamic lot sizing model embedded in a scenario approach and solved by MIP solver

continued

Table 5.3: Overview of related literature part III: Frequency assignment with delivery profiles (continued)

Author	Considered cost	Considered transport concepts	Decision on frequency	Solution approach
Harks et al. (2014)	group of complex tariff structures and inventory holding cost	(1) groupage service with different tariff tables on links of pre-leg, main-leg and sub-leg (2) point-to-point transport	cyclic pattern (with a cycle time of 2 months in the case study)	pattern expanded multi commodity capacitated network design formulation, initial solution by heuristic relying on shortest path augmentations and LP techniques, improvement phase by local search rerouting flow of multiple commodities at once combined with aggregated mixed integer solutions attained by a MIP solver

The improved approach of (Schöneberg et al. 2010) was evaluated on six representative areas of a German truck manufacturer considering 3 to 16 weeks. As a benchmark, they use solutions of the material resource planning component not considering transport cost at all. Over all scenarios the cost savings realised are between 0% and 37% with an average of 17%. It is interesting to see that the inventory cost increased on average by only 4% with a maximum of 7%.

In (Schöneberg et al. 2013) a simplified version of the model is extended to a two stage stochastic model. It is embedded in a scenario based approach to determine delivery profiles robust towards stochastic demands and frequencies. The solution framework consists of scenario reduction techniques, a decomposition technique, a genetic algorithm and a MIP solver. The approach was also evaluated on real-world data containing 25 areas of an area forwarding network. On average, the savings reached 23% using the deterministic approach and could be enlarged by further 3 percentage points using the robust version.

In the work of Meyer et al. (2011) the goal of introducing delivery profiles – called transportation pattern – is different: The clients of a freight forwarding network choose a set of delivery profiles and leave it to the freight forwarder to choose one of the profiles in order to achieve consolidation and levelling effects on the network level considering the orders of all clients.

Figure 5.5: Generation of a delivery schedule by applying a delivery profile with delivery days Monday, Tuesday, Thursday to a net demand forecast using the dynamic lot sizing model of (Meyer et al. 2011)

The first stage problem is modelled as a dynamic lot-sizing problem with additional constraints to level out the transport lots over time. This step is embedded in a scenario approach in order to generate robust solutions in terms of stochastic demands. The second stage is modelled as a tactical network planning problem not explicitly considering routing.

In contrast to the preceding approaches, on the operational level, a model similar to the one of the first stage tactical model is used to generate delivery schedules with levelled lot sizes considering the transport pattern, which was assigned by the freight forwarder (see Figure 5.5 showing the net demand over the respective due date on the left and the resulting transport lots in the right chart).

The approach was evaluated on a big real-world network with 3500 shipping relations and 10 hubs. It could be shown that the number of pick up and delivery tours could be reduced by around 7% and, at the same time, the use of resources could be levelled out over the course of the week compared to a solution without applying the levelling mechanisms.

Harks et al. (2014) also propose the introduction of delivery profiles – called cyclic patterns – in order to be able to realise spatial and temporal consolidation effects in a logistics network. They proposed a multi commodity network design model considering capacities and different groups of complex tariff structures. The network model is expanded by cyclic patterns. The goal is to assign a cyclic pattern to each order and to define a path through the transport network, which consists, for example, of groupage services with different tariff tables on pre-, main- and sub-legs and direct relations.

The solution procedure contains different heuristics, relying on shortest path augmentation and linear programming (LP) techniques for generating an initial solution and local search steps, which re-route flows of multiple commodities at once, in an improvement phase. The evaluation of the approach on a big real-world instance with 228 consumers, 545 suppliers and 5 hubs yielded a 14% cost reduction compared to a solution from practice.

5.2.2 Pull Based Kanban Systems

Similarly to the preceding section for push based order policies, we differentiate in case of the pull based kanban system between literature assuming a constant cycle and literature considering differing inter arrival times levelled by heijunka mechanisms.

Kanban Systems Neglecting the safety stock part, the Toyota formula introduced in Section 3.4 yields the minimum number of kanban for fixed cycle and lead times and a given kanban size. That means the parameters $a - b - c$ and the kanban size M are given. However, in a first step obviously the delivery cycle time $\frac{a}{b}$ and the kanban size M have to be determined.

All together the number of kanban, the cycle time or order interval, the kanban size, and the safety stock level are considered as design parameters of a kanban system. The goal when determining these parameters is usually to minimize inventory, to maximize the throughput or to minimize operational cost including holding cost, shortage cost, and setup cost (see (Akturk and Erhun 1999)).

Common methods to determine the design parameters proposed in literature comprise simulation based models, mathematical programming models, and queuing and Markov chain models (Matzka et al. 2012). A good overview and classification of the early literature is given in (Akturk and Erhun 1999), a more recent review is given in (Kumar and Panneerselvam 2007).

Miyazaki et al. (1988), for example, determine optimal order intervals and the optimal number of kanban for a supplier kanban system in a deterministic environment considering inventory holding and a fixed setup cost. The authors show that the difference compared to the EOQ model lies in a different shape of the cost curve: The curve in the kanban model is not a convex function as in the EOQ model.

In (Kotani 2007) the special case of supplier eKanban is treated and methods to optimally adapt the number of kanban under changed conditions are proposed. The work relies on the classic Toyota Formula. A more detailed introduction to a basic inventory model for kanban with constant cycles follows in Chapter 7.

Heijunka Levelled Kanban Systems It is also Kotani (2007) who states that the time intervals between two deliveries must be constant to the greatest possible extent. This is difficult to guarantee, for example, due to differing operating hours of suppliers and plants. It is even more difficult in case of weekly milk runs, in which fixed inter arrival times lead to strongly varying arrival times at the supplier (remember Figure 3.5) which are not in line with opening hours or legal night drive bans. For these cases, it is proposed by (Chuah and Yingling 2005) and (Ohlmann et al. 2008) to apply heijunka levelling in order to achieve transport lots of equal size.

However, neither Chuah and Yingling (2005) nor Ohlmann et al. (2008) introduce inventory models for these cases. Chuah and Yingling (2005) approximate inventory cost ignoring different inter arrival times, while Ohlmann et al. (2008) neglect inventory cost. Both approaches seem sensible in case of milk runs executed several times per day because routing cost – determined exactly by solving a VRP (see next section) – usually exceeds holding cost significantly.

To the best of our knowledge there exist only a few models to analyse and dimension heijunka systems, such as (Lippolt and Furmans 2008), (Furmans and Veit 2013), and (Matzka et al. 2012). They all focus on buffer sizing in production systems considering stochastic processes. None of the listed approaches assumes that the time between two shipments is varying purposefully as it is the case if weekly inter arrival time consistent schedules are applied for milk runs.

5.3 Milk Run Scheduling

The milk run scheduling decision described in Section 4.3 contains a vehicle routing decision. Vehicle routing problems (VRP) are one of the most studied problem classes in operations research and it is not useful to give an extensive overview of all classes of VRP related to the problem.

However, for a better understanding of the approaches addressing milk run planning or specific aspects important for milk run planning, we start with a short introduction of basic VRP classes.

5.3.1 A Brief Introduction of Related VRP

For a better overview of VRP classes related to the Milk Run Design, we give typical characteristics of the most important classes discussed in this section in Table 5.4. The characteristics are differentiated by the scope of the decision, by typical constraints, and by typical cost considered in the objective functions. The starting point is the basic capacitated vehicle routing problem (CVRP).

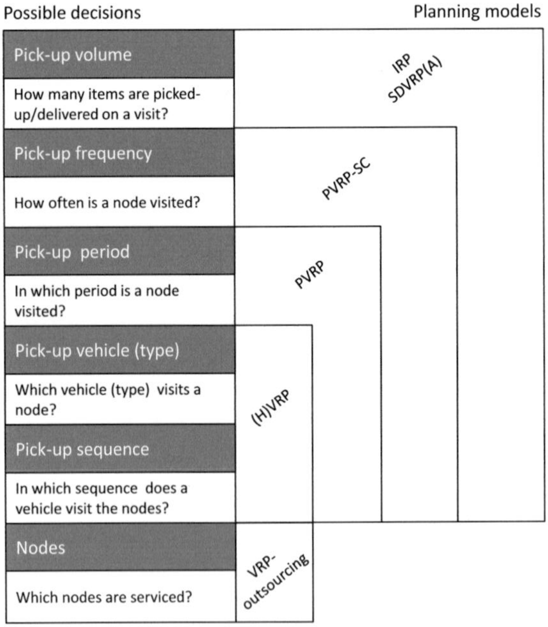

Figure 5.6: Scope of decision in vehicle routing problems and the corresponding planning models

Table 5.4: Typical characteristics of relevant vehicle routing models

		CVRP	PVRP	PVRP-SC	IRP	SDVRP	VRP & Outsourcing	conVRP	PVRP & Complexity	HVRR
Decision	nodes	–	–	–	–	–	✓	–	–	–
	sequence	✓	✓	✓	✓	✓	✓	✓	✓	✓
	vehicle	✓	✓	✓	✓	✓	✓	✓	✓	✓
	period	–	✓	✓	✓	–	–	–	✓	–
	frequency	–	–	✓	✓	✓	–	–	–	–
	volume	–	–	–	✓	✓	–	–	–	–
	vehicle type	–	–	–	–	–	–	–	–	✓
Constraints	vehicle capacity	✓	✓	✓	✓	✓	✓	✓	✓	✓
	storage capacity	–	–	–	✓	–	–	–	–	–
	arrival time cons.	–	–	–	–	–	–	✓	–	–
	driver consistency	–	–	–	–	–	–	✓	–	–
Objective	cost types	T	T	T S	T H	T	T O	T	T C	T

T	variable (and fixed) transport cost
S	service benefit
H	inventory holding cost
O	outsourcing cost
C	penalty cost for complexity

Figure 5.6 illustrates what we mean by scope of the decision, and how it is related to different model classes. We restrict ourselves to the inbound case considering pick ups from suppliers transported to a receiving plant[1]. However, in case of a distribution problem, all decisions are analogue to deliveries.

[1] In the following we refer to the suppliers and the receiving plant of a VRP as nodes.

The Periodic Vehicle Routing Problem Decisions (3) to (5) of the Milk Run Design introduced in Section 4.4.1 are: (3) assigning every supplier to periods considering feasible patterns, (4) assigning every visit to a vehicle available in the corresponding period, and (5) determining the sequence of visits for every vehicle. These decisions typically describe periodic vehicle routing problems (PVRP). In Figure 5.6 they are represented by pick-up sequence, pick-up vehicle and pick-up period. An overview of this problem class and its extensions is given in (Francis et al. 2008)).

If the frequency is not fixed in advance, the PVRP is extended to a PVRP with service choice (PVRP-SC introduced in (Francis et al. 2006b)), in which a minimum service frequency is guaranteed. This lower bound is either exceeded in order to obtain compatible volumes for neighbouring nodes and to save an extra trip to this cluster of neighbours, or a service benefit in the objective function compensates for extra travel costs. In our case, this service benefit corresponds to inventory cost savings.

In PVRP models the shipping volumes are typically assumed to be equal for all shipments: either the demand for the planning horizon is assumed to be divided in shipments of equal size. Or an upper bound for the shipment size is used in the model derived by multiplying the demand rate with the maximum allowed inter arrival time (cmp. for example (Francis et al. 2006b)) – even if the actual amount is the demand accumulated since the last visit or until the next. Obviously, in such a case consolidation potential is wasted.

The Split Delivery Vehicle Routing Problem Besides the PVRP-SC there exists a second problem class in which the number of services at one node is a decision variable: the vehicle routing problem with split delivery (SDVRP). For a literature review we refer to (Archetti and Speranza 2008). Typically, this problem variant is not periodic, i.e. it does not contain a decision about the period in which a visit is performed. Rather, one assumes that more than one vehicle can visit a customer and the shipping amount is split in a way that the vehicle capacity is optimally used. A periodic version (SDVRPA), where also a pick-up period is determined, was recently introduced by (Breier and Gossler 2014). The way how the demand of a client is split into lots is usually not restricted.

The Inventory Routing Problem If the inventory levels and the corresponding inventory holding cost for each time unit on stock – possibly including penalty cost for stock outs – are explicitly considered, the model is, usually, classified as an inventory routing problem (IRP see also Figure 5.6). Typically, there is no minimum service frequency imposed, but sometimes the inventory levels are capacitated. The time of delivery is unconstrained and the shipping amounts are either free – i.e. purely defined by optimal inventory and routing cost – or defined by an order policy such as an order-up-to-policy. For a profound introduction the reader is referred to (Campbell et al. 1998) and (Andersson et al. 2010).

Considering Outsourcing in VRP A common operational problem of a freight forwarder is to decide which part of the daily orders should be fulfilled with vehicles from the own fleet, and which part should be outsourced to subcontractors by minimizing the sum of fixed cost of used vehicles, travel and subcontracting cost. This decision problem is referred to as prize collecting vehicle routing problem in (Stenger et al. 2013), while Wang (2015) introduces it as vehicle routing and forwarding problem and a more general variant as integrated operational transportation planning problem.

Both (Stenger et al. 2013) and (Wang 2015) give an extensive overview of literature and propose mathematical models and heuristics covering, amongst others, the outsourcing to a so called common carrier. The latter is to a freight forwarder what a groupage service provider is to a producing company. Usually, the freight forwarders also arrange medium-term contracts regulating the tariffs with these carriers.

Another important aspect when outsourcing tours is that the vehicles do not necessarily start and end at known depots. If – instead of round tours – such a tour is planned, the VRP is classified as open vehicle routing problem (OVRP), in which a route is defined as a Hamiltonian path over the subset of visited customers. An overview of the problem class and corresponding algorithms is given, for example, in (Li et al. 2007). Furthermore, in practice different subcontractors offer different vehicle types and tariffs: If more than one vehicle type with distinct capacity or cost parameters is available, the problem is classified as heterogeneous vehicle routing problem (HVRP) or fleet size and mix vehicle routing problem (for an overview see (Baldacci et al. 2008)).

Considering Consistency in VRP In (Coelho et al. 2012) different types of consistency are introduced for the IRP in order to increase the quality of service in vendor managed inventory systems compared to solutions purely optimizing routing and inventory cost. Relevant consistency concepts are the following three (see also Section 2.4): The quantity consistency enforces the shipping amounts to lie within an interval around a target value. The driver consistency ensures that exactly one vehicle is assigned to a customer over the planning horizon, whereas the partial driver consistency minimizes the number of assigned vehicles. In both cases, a fixed connection between vehicle and driver is assumed. The visit spacing requirement in (Coelho et al. 2012) assures that at least every M periods a visit will take place, and no more than one visit will take place in m successive periods. Service times within a day – as in the described inter arrival time consistent (ITC) or arrival time consistent (ATC) schedules – are not considered.

Arrival time consistent schedules have been introduced in (Groër et al. 2009) in the context of package delivery: Groër et al. (2009) define the consistent vehicle routing problem (ConVRP), in which the same customers are serviced by the same driver at roughly the same time of the day over the planning horizon in order to stimulate a good relationship of drivers with frequent customers and provide predictable arrival times. The days of service in this model are known in advance. Related concepts have been introduced before by Francis et al. (2006a) for the PVRP, with the difference that the number of driver changes and the arrival span (maximum time between two arrivals on different days of service) are not added as restrictions, but are minimized in order to reduce the "operational complexity" of a schedule. Besides the so called customer familiarity of the driver, also a region familiarity is considered (see also an extended study in (Smilowitz et al. 2013)).

Similar to the arrival time consistent VRP is the time window assignment vehicle routing problem (TWAVRP) introduced in (Spliet 2013), for example for distribution networks in retail. The assignment of a time window from a set of feasible time windows for clients is a long-term decision without knowing the concrete demand realisation. The assignment is done in a way that the resulting operative tours are cost-efficient with guaranteed arrival times for the customers. This approach resembles the master tour concept of Section 2.1.2. A good overview of further literature proposing slightly different concepts of consistency in VRP is given in (Kovacs et al. 2014).

Considering Synchronization in VRP All consistency or complexity reduction requirements as well as the consideration of inventory cause an interdependence or a so called synchronisation between tours and, therefore, can be subsumed under the class of VRP with synchronisation introduced by (Drexl 2012). That means, for example: The time window of a second visit depends on the start of service of a first visit, and, hence, a change of one tour affects other tours. Either the contribution to the objective value of another tour might be changed or, in the worst case, another tour becomes infeasible (see (Drexl 2012)). In case of inventory routing, the timing of one visit affects the loads of a later visit and, hence, a later tour.

5.3.2 VRP Related to Modelling Milk Run Scheduling

In the following section we focus on literature considering milk run systems and very similar concepts for in- and outbound processes. For an overview and classification scheme of in-plant milk runs the reader is referred to (Kilic et al. 2012). As shown in Figure 5.7, we differentiate again between models resulting in constant delivery cycles and models resulting in delivery profiles repeated over time for every supplier.

Constant Cycles As already mentioned in Section 3.5, some authors introduced models directly addressing the route design decisions for the in-bound milk run system of lean production systems: Chuah and Yingling (2005) introduced the common frequency routing (CFR) policy combining only suppliers with the same frequency on one tour, while Ohlmann et al. (2008) allowed that suppliers with differing frequencies are combined on the same tours. The latter is called general frequency routing (GFR).

We differentiate the models proposed in literature by the following three features (see Figure 5.7):

- **CC:** All considered suppliers of the network have the same constant cycle time or inter arrival time and, hence, the same frequency. They can be combined on different tours which all have the same frequency.
- **CC-T (FPP):** All suppliers on the same tour have the same constant cycle time. That means, only suppliers with the same frequency are combined on the same tour. In literature, this is also referred to as fixed partitioning policy (FPP) (see for example (Ekici et al. 2014)).
- **CC-S:** Every supplier has its own constant cycle time, and suppliers of different frequencies can be combined on the same tour.

We furthermore describe the basic models and the relevant extensions and consistency features, the length of the planning horizon and the corresponding base period. Solution approaches are introduced only briefly. Furthermore, we shortly describe the application area differentiated by inbound (IN) and outbound (OUT) use cases. Again, we only present a selection of the most relevant literature listed and described in Table 5.5. Especially, we do not include work that considers only one base period.

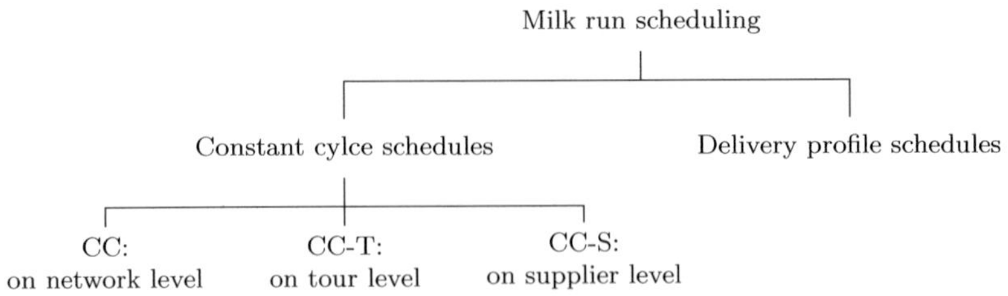

Figure 5.7: Top level classification scheme of literature related to milk run scheduling

Table 5.5: Overview of related literature part IV: Milk run scheduling with constant cycles

Author	Model type	Planning horizon, Base period	Lot sizing policy	Consistency	Solution approach	Use case
Chuah and Yingling (2005)	CFR: SDVRP with TW, frequencies restricted to $\{1, 2, 3, 4, 8\}$	1 day, 3 hours	equal split (HEIJUNKA)	CC-T: CFR, explicitly only scheduling the first visit, assuming that an appropriate spacing of visits is possible (restricted by TW), driver consistency possible	Column Generation or tabu search for solving the SDVRP, scheduling the tours equally distributed over the day	IN: Toyota, USA
Stacey et al. (2007), Natarajarathinam et al. (2012)	IRP restricted to constant order intervals, restricted number of clients per tour	infinite, result of optimization	equal split	CC-T, driver consistency possible	improved approaches of Natarajarathinam et al. (2012): lower bound based on Lagrangian relaxation, two heuristics: (1) iterative routing phase and frequency phase, (2) iteratively executing both phases	IN: automotive industry
Grunewald et al. (2011)	IRP with stochastic demand with levelling order policy and chance constraints	1 day, 1 day	control band policy	CC: daily the same tours, driver consistency possible	MIP solver for a small example	IN: automotive industry

continued

Table 5.5: Overview of related literature part IV: Milk run scheduling with constant cycles (continued)

Author	Model type	Planning horizon, Base period	Lot sizing policy	Consistency	Solution approach	Use case
Ohlmann et al. (2008)	GFR: SDVRP with TW	1 day, several hours	equal split (HEIJUNKA)	CC-S: ITC as weighted objective in the scheduling phase (restricted by TW)	2-stage heuristic: nested tabu search for routing and MIP for scheduling	IN: Toyota, USA
Raa and Aghezzaf (2009)	Cyclic IRP with multiple use of vehicles and frequency restriction to power-of-two	infinite, 1 day	equal split	CC-T: tours always include the same customers with a common frequency, but one vehicle can take over different tours with different frequencies, driver consistency possible	solution framework contains construction heuristics, local search, set cover based heuristic, greedy scheduling algorithms	OUT: VMI
Chen and Sarker (2014)	MVSB-VRP: production lot sizes for suppliers, economic transport lot sizes, routes and delivery frequencies	infinite, result of optimization	equal split	CC: all suppliers have the same constant cylce time, driver consistency possible	ant colony algorithm with sub algorithms for MVSB and VRP	IN: JIT systems

Chuah and Yingling (2005) and Ohlmann et al. (2008) both propose models for planning milk run systems with a base period of several hours and a system cycle time of one day, i.e. the milk run schedules are repeated in the same way every day. Both models can handle the case in which perfectly equal arrival times cannot be reached due to differing opening hours of the suppliers and the plant: The deviation of the optimal inter arrival times are considered as a part of the objective function.

In both papers, the base model is a split delivery VRP with time windows in which the lot size decision is restricted to an equal split of the daily demand. The latter is assured by levelling in case of differing inter arrival times. Chuah and Yingling (2005) argue that CFR is applied to reduce the operational complexity since all suppliers are always serviced on the same milk run and these routes are consistently repeated over time. According to the authors, this restriction leads only to small increases in transportation cost but has the positive effect of reducing the dimensionality of the search space.

Ohlmann et al. (2008) omit this restriction since, according to their results, a significant cost saving of up to 7% for the example instance introduced in (Chuah and Yingling 2005) can be realised and a better 'workload levelling' is possible. That means, one reaches a schedule in which the arrival times at the suppliers are distributed more equally over the operating hours of the suppliers. Furthermore, in contrast to Chuah and Yingling (2005), Ohlmann et al. (2008) also consider the period between the last visit of a day and the first one of the next day.

The goal of maximizing the consistency of inter arrival times corresponds to the goal to reach a low WIP. However, this relationship has not been mentioned in either of the work. Ohlmann et al. (2008) completely ignore inventory holding costs since – if they are "traditionally calculated" – they are very small compared to the routing cost. In (Chuah and Yingling 2005), they are estimated on frequency level but only account for less than 2% of the objective function value in the real-world example.

From these results we conclude that choosing a high frequency for a supplier is either driven by routing costs – obviously, much more important in the CFR case, in which neighbouring suppliers should be assigned to a common frequency – or by the storing capacity restrictions. The latter is realised by capping the theoretical peak inventory summing up the total shipment sizes of all suppliers. For a more detailed consideration of the space consumption of the shipments in the receiving plants it would be necessary to model inventory balance over the course of the day. Hence, the capacity restriction is rather interpreted as a WIP-cap, restricting the total work in progress, which is common in lean manufacturing. The actual size of the WIP-cap is set by the decision maker.

Chuah and Yingling (2005) present a column generation approach for solving small problems and a tabu search for solving the frequency assignment and routing problem of real-world instances with up to 54 orders. During the tabu search, only the first visit of every supplier is scheduled explicitly, assuring that it is early enough so that the later tours can be scheduled in the remaining time. The concrete scheduling of the tour starts is done as a second step by distributing them equally.

Ohlmann et al. (2008) also propose a two stage approach: a nested tabu search iteratively determining frequencies and vehicle routes for the routing part and a binary integer program minimizing the deviations from the optimal inter arrival times for all suppliers simultaneously for determining the schedules. The approach is evaluated on real-world instances with up to 100 suppliers.

Stacey et al. (2007) propose a very similar model to the one in (Chuah and Yingling 2005) but without considering opening hours: Routes contain only suppliers of the same frequency, and the routes are scheduled according to the optimal cycle time calculated by using the extended EOQ formula of (Blumenfeld et al. 1985) on route basis. In (Natarajarathinam et al. 2012), the same authors propose improved heuristics and a lower bound based on Lagrangian relaxation for the nonlinear problem.

Since the planning horizon is infinite and the cycle times per tour are not restricted to a few frequencies or to integer periods (i.e. one day), the optimal cycle time is most probably different for every tour. Therefore, the time until the whole tour plan is repeated for the first time – the system cycle time – can be long. It corresponds to the greatest common multiple of all tour cycles. However, a single tour is repeated always after a fixed interval so that driver consistency can be easily implemented.

In (Raa and Aghezzaf 2009) – based on the preceding work (Raa and Aghezzaf 2008) – a cyclic replenishment policy for a vendor managed inventory system (VMI) is proposed in order to provide stable and predictable plans to the customers. The authors also restrict the model in a way that only suppliers of the same frequency can be served on the same tour. However, they allow multiple use of vehicles per day, and the same vehicle can be assigned to tours of different frequencies. In contrast to (Stacey et al. 2007), the base period is restricted to one day. The solution framework of (Raa and Aghezzaf 2009) contains several building blocks, such as construction heuristics, local search, a set cover based heuristic and greedy scheduling algorithms.

In a very recent work, also the production lot sizing problem at the supplier sites is included: Chen and Sarker (2014) propose a model combining the optimal production and transport lot sizing problem with a vehicle routing problem to a so called multi supplier single assembler system with JIT delivery. In literature, such a system is also classified as multi vendor single buyer problem (MVSB). That means, the approach determines in an integrated fashion the production lot sizes for suppliers, economic transport lot sizes, milk run routes and delivery frequencies. The model is restricted to the CC case, that means, all suppliers and tours have the same frequency. The planning horizon is infinite and the cycle time is not restricted to an integer multiple of a base period. For solving the problem an ant colony based meta heuristic and sub algorithms for the MVSB and VRP are proposed.

Another work treats production and purchasing decisions using milk runs as transport concept. An approach for small examples considering a power-of-two policy is given in (Kuhn and Liske 2014). However, since the coordination of production and transport planning is beyond the scope of this work, the reader is referred to an overview of the related work recently given by (Kuhn and Liske 2014).

A special case of milk runs with constant cycles is described in (Grunewald et al. 2011). The authors assume stochastic demands for parts which are picked-up every day: They propose a stochastic variant of the IRP imposing a chance constraint on the vehicle capacity. This means that the capacity restriction of the vehicle must hold at least in a certain percentage of the scenarios, which are generated by simulation. This is similar to the very common service level concept in logistics. In addition, the authors seek for levelled inventory and similar order volumes. Therefore, a so called order bandwidth policy is introduced to keep the shipping amounts within an interval around the mean demand. This range is – besides the daily routes – an output of the optimization process depending on inventory and transportation cost. Please note that such a policy is similar to heijunka levelling, in which the maximum delivery amount is capped. In (Grunewald et al. 2011), only the result of a very small instance was discussed without proposing a solution procedure for real-world instances.

Delivery Profiles Approaches to determine regular tours – i.e. milk runs – resulting in regular delivery profiles for every customer have been first proposed in literature for outbound networks in food industries (see Table 5.6).

Gaur and Fisher (2004), for example, introduce an approach to determine weekly recurring tours for supplying the stores of Albert Heijn, a super market chain in the Netherlands. For reducing operational complexity and the search space, stores are always serviced in groups: All stores are either serviced on a shared tour with all supermarkets from its cluster or on direct tours. This leads to similar tours over the course of the week, which can be serviced by the same driver or a group of drivers. The underlying model is a periodic inventory routing problem, in which the shipment size is determined by the demand accumulated until the next scheduled visit. On the last visit of the week, the demand until the first visit of the next week is accumulated. The inter arrival time is limited by a maximum interval between two consecutive visits, which is imposed due to perishable goods. For solving the problem, clusters of stores are built heuristically, and a shortest path heuristic is used to solve a simplified inventory routing problem for every cluster. Subsequently, a weighted matching heuristic selects the best subset of clusters. In the end, a scheduling fixes the start times of tours in a way that the workload in the distribution centre for loading and unloading the trucks is balanced.

Table 5.6: Overview of related literature part V: Milk run scheduling with delivery profiles

Author	Model type	Length of horizon	Volume split policy	Consistency	Solution approach	Use case
Gaur and Fisher (2004)	periodic IRP	1 week, 1 day or fraction of 1 day	demand accumulated to the next shipment	groups of stores are replenished together, either in a direct route or a shared route, schedules balance workloads at depot	shortest path heuristics to solve inventory routing problems for clusters of stores, matching heuristic to choose the best clusters, truck assignment and scheduling heuristics	OUT: Albert Heijn, Netherlands
Rusdiansyah and Tsao (2005)	PVRP-SC-TW extended by inventory model, frequencies restricted to $\{1, 2, 3, 6\}$	6 days, 1 day	equal split due to restricted frequencies	CC-S: constant cycles due to restricted frequencies, ATC guaranteed by pre defined TW	insertion based initialization followed by tabu search	OB: vending machine refilling, Japan

continued

Table 5.6: Overview of related literature part V: Milk run scheduling with delivery profiles (continued)

Author	Model type	Length of horizon	Volume split policy	Consistency	Solution approach	Use case
Parthanadee and Logendran (2006)	multi depot PVRP-SC considering inventory levels and backorders	1 week, 1 day	lot size corresponds to demand until next delivery	“well spaced” weekly pattern, customer-defined preferences for patterns considered in the objective function by penalty cost	insertion based construction heuristic with local search on routes, tabu search for changing depot allocation and pattern assignment	OUT: SYSCO USA
Alegre et al. (2007)	PVRP	up to 90 days, 1 day	not reported, assumed as equal splits	“well distributed” pick ups, no repetitive tours	Scatter Search	IN: Bentzeler, Spain
Esparcia-Alcázar et al. (2009)	PVRP-SC considering inventory	1 week, 1 day	equal split	“well spaced” delivery pattern	evolutionary algorithm to assign pattern and savings enhanced by local search for VRP	OUT: supermarket distribution, Spain

continued

Table 5.6: Overview of related literature part V: Milk run scheduling with delivery profiles (continued)

Author	Model type	Length of horizon	Volume split policy	Consistency	Solution approach	Use case
Peterson et al. (2010)	SDVRP with stochastic demand handled by (I) chance constraints (II) a recourse strategy	1 week, 1 day	free choice of volumes restricted to integer multiples of a minimum lot size	no explicit scheduling of the routes	a priori column generation and set cover restricting the number of visits per tour to 3	IN: BSH, North America
Sadjadi et al. (2009), Jafari-Eskandari et al. (2010)	IRP with multi sourcing, stochastic inventory (2nd paper)	1 week, 1 day	free choice of volumes	none	MIP, Genetic Algorithm, Particle Swarm	IN: Saipa, Iran
Ekici et al. (2014)	Cyclic IRP avoiding stock out but not considering inventory cost	T days or submultiple of T days, 1 day	free choice	none	iterative cluster generation and delivery schedule generation heuristic	OUT: industr. gas manufacturer

Rusdiansyah and Tsao (2005) solve a very similar problem for refilling vending machines: Their basic model is a PVRP with service choice and time windows extended by an inventory model. The authors restrict the frequencies to divisors of the planning horizon of 6 days, i.e. to 1, 2, 3 and 6. By doing this and by considering time windows for the start of service at every vending machine, the resulting schedules are, automatically, arrival time and inter arrival time consistent (remember the example of Figure 3.5) and, hence, fulfil a CC-S policy. But, for a working week of 5 or 7 days this model seems to be inadequate due to only 2 feasible frequencies. The proposed solution approach is an insertion based starting heuristic followed by a tabu search.

In the work of (Parthanadee and Logendran 2006), a multi depot variant of a food distribution application is considered: That means, the customers can be served from different depots with restricted capacity, and the goal is to minimize transport and backorder cost. The proposed model is a multi depot periodic vehicle routing problem allowing only "well spaced" patterns – that means, for example, no consecutive delivery days in case of a frequency of two per week. Furthermore, clients can give priorities to patterns, and a component of the objective function value incentivises the assignment by applying penalties to the less preferred ones. The lot sizes simply correspond to the demand until the next scheduled delivery. As solution approach, the authors propose an insertion based construction heuristic enhanced by local search steps on route level and a tabu search for changing the depot allocation and pattern assignment.

The PVRP-SC model introduced in (Esparcia-Alcázar et al. 2009) is meant for generating a weekly recurring tour plan for a supermarket chain in Spain. It considers inventory and transport cost. The delivery patterns are also restricted to the "well spaced" ones. A minimum frequency, and hence a feasible set of patterns, can be determined for every client by considering, for example, the space restrictions at the stores. It is assumed that the weekly demand is equally split into transport lots and that an estimation for the resulting inventory cost is given on frequency level. However, no explicit inventory model is introduced. For solving this problem, an evolutionary algorithm assigning one of the feasible patterns to every client and a savings heuristic enhanced by local search for the resulting VRP are proposed.

For an inbound use case of Bosch Siemens Home Appliances (BSH) in North America, Peterson et al. (2010) propose so called “flexible milk runs” or “flex runs”. In spite of correlated stochastic demands, these flex runs shall offer weekly recurring fixed schedules in order to have stable and predictable distribution patterns. The authors propose two techniques to achieve this goal: (I) They introduce a chance constraint formulation as described in the preceding section (see (Grunewald et al. 2011)), and (II) they propose a 2-stage stochastic model optimizing the cost for the actual flex runs along with repair cost in case of exceeding capacities for a given recourse (repair) strategy. The underlying VRP model is a SDVRP with free choice of volume splits not explicitly considering the assignments on days of the week. However, after solving the SDVRP the routes can be easily assigned to week days, and for the suppliers a delivery profile is formed.

For conducting the study, the authors proposed a set covering based heuristic: They generated all routes with up to 3 suppliers and for all these routes all meaningful combinations of the pick up proportions – every proportion must be a multiple of the minimum lot size – resulting in a set of around 600,000 flex run candidates for the real-world instance of 75 suppliers. The feasibility with respect to the chance constraint can be checked, and the cost for applying a recourse action can be calculated on candidate level. A set cover is solved with an integer programming solver to choose the best flex runs out of this candidate list. The resulting milk run schedule is repeated weekly, but within the week there are no further requirements on regularity. With this approach, savings of 25% could be achieved for the considered network.

Sadjadi et al. (2009) propose an IRP model for an Iranian automotive manufacturer which seeks for weekly recurring schedules. Beyond that, there are no requirements on regularity, i.e. volumes and schedules can be freely determined over the course of the week. To solve the model a MIP formulation and a genetic algorithm are proposed, for the stochastic variant a robust MIP formulation is solved with an integer programming solver for a small example.

As Raa and Aghezzaf (2009) mentioned in the preceding section, Ekici et al. (2014) propose a cyclic IRP for the distribution of an industrial gas manufacturer, however, with a different understanding: A fixed planning horizon of T days is given. The clients' inventory at the beginning of T and at the end of T are equal so that the tours can be repeated again and again. Hence, the planning horizon T corresponds to the system cycle time. Within this horizon, the deliveries are freely assigned to days, and the transport lots are chosen in a way that no backlogs occur. Inventory costs are neglected. The problem is solved for the system cycle time of T and for all submultiples: A feasible plan for 2, 7 and 14 days results in a cyclic plan for 14 days. The authors propose an iterative cluster and delivery schedule generation for solving this problem.

A special case is the approach of Alegre et al. (2007), who solve a classic PVRP model by a scatter search based heuristic for the periodic pick up of materials for a long-term planning horizon of up to 90 days. The approach is proposed for the inbound logistics of Bentzeler, a supplier for car manufacturers in Spain, assuming that the production plans are fixed for the planning horizon. The regularity is reached on supplier level by allowing only pick up calendars with equally spaced pick up dates, however, a repetitive schedule in terms of tours is not the goal.

5.4 Summary

In this chapter we showed that the literature related to modelling the Milk Run Design decision is diverse. We classified the approaches by examining which transport concepts and cost structures are considered, and which types of regularity and consistency the resulting schedules show.

The transport concept assignment can be split in an operational and a tactical task. The operational decision is modelled as network flow or VRP model or as a combination of both. Only in one approach delivery profiles are considered, but none of them considers milk runs as a transport concept.

The tactical transport concept decision is combined with the frequency assignment. This problem is differentiated by push based order policies, pull based kanban systems, by approaches resulting in schedules with constant cycles and by approaches resulting in schedules with delivery patterns or profiles.

In case of push based policies for both constant cycle and delivery profile models, there exist approaches to determine the optimal frequency considering complex tariff structures as they exist for groupage services. In case of kanban systems, only constant setup cost has been considered for determining the optimal cycle time.

Models determining milk runs are also differentiated by the type of regularity of their schedules: constant cycle approaches, further divided into CC, CC-T, and CC-S, and delivery profile approaches. Appropriate approaches have been proposed in the literature which simultaneously assign a frequency or delivery profile to every supplier and determine the milk run routes for all types. To this end, basically three different VRP classes are used: IRP, SDVRP, and PVRP-SC.

An evaluation of the proposed models with respect to the Milk Run Design decision follows in the next chapter.

6 Research Gaps and Opportunities

In the light of the insights of Chapter 2, 3, and 4, we describe the strengths and gaps of the related models existing in literature with respect to the Milk Run Design decision and derive the need for action. Furthermore, we describe how these issues are addressed in the remainder of this thesis.

6.1 Strengths and Weaknesses of Existing Models

The literature overview of Chapter 5 shows that there exist promising models and solution approaches addressing one or two of the tasks (1) transport concept assignment, (2) frequency assignment, and (3) milk run scheduling resulting from the decomposition of the Milk Run Design problem introduced in Chapter 4. However, none of the models considers the fully integrated problem. Furthermore, not all aspects possibly important in a milk run inbound network are considered. In the following we assess the existing models in order to derive extended variants addressing the Milk Run Design problem.

Constant Cycle Milk Run Scheduling In our judgement, especially the models of Chuah and Yingling (2005) and Ohlmann et al. (2008) are well suited for planning daily recurring milk run schedules: Only in these works the complicating aspect of varying opening hours of the suppliers and the plants – this situation is the rule rather than an exception – is addressed. However, the inventory behaviour at the plant site and the supplier site in case of purposely planned varying inter arrival times combined with heijunka levelling is not investigated.

If opening hours play a minor role, the models presented in (Stacey et al. 2007) and (Natarajarathinam et al. 2012) are of interest. Whenever additionally the multiple use of vehicles due to short round tours provides savings potential, the approach of (Raa and Aghezzaf 2009) might be suitable. However, both models assume an infinite planning horizon and do not seek for repetitive schedules but only for repetitive tours. This results in overall complex schedules with varying cycle times for tours and potentially long system cycle times. The use of resources during this system cycle time cannot be controlled with these approaches.

The models of Grunewald et al. (2011) and Chen and Sarker (2014) consider interesting extensions – stochastic demands and the integration of the production lot sizing problem at the suppliers respectively – but both are restricted to the CC case. The applicability in practice is therefore limited since the number of suppliers which can be served cost efficiently by tours with the same frequency is usually very small since volumes and values of parts differ considerably.

Delivery Profile Milk Run Scheduling Rusdiansyah and Tsao (2005) propose a model for weekly recurring milk runs which is limited to feasible frequencies $\{1, 2, 3, 6\}$ and a system cycle time of 6 days. For system cycle times which are not a multiple of two – for example a 5 day working week – it is not applicable. Furthermore, it is assumed that the time windows are fixed in advance. However, in the milk run use case in practice time windows can be negotiated and should be a result of the decision process in order not to lose optimization potential. In the model of Alegre et al. (2007), the focus lies on generating schedules with a long-term validity of up to 90 days for the suppliers. A generation of repetitive schedules is not considered.

The models of Peterson et al. (2010), Sadjadi et al. (2009), and Ekici et al. (2014) all result in weekly or bi-weekly (only (Ekici et al. 2014)) recurring milk run schedules. The transport lot sizes of a supplier over the week are chosen in a way that the vehicle capacity is ideally used. Only Sadjadi et al. (2009) additionally consider inventory cost. Furthermore, in all three models the shipment days can be freely chosen: In an extreme case, in an optimal schedule a shipment on Monday satisfies 90% of the weekly demand and one on Tuesday the remaining 10%. There exist use cases in practice in which an unlevelled transport schedule yielding small transport cost is beneficial. However, such a policy does not promote a smooth flow of parts and is not in line with lean principles. An interesting aspect of the work in (Jafari-Eskandari et al. 2010) and (Peterson et al. 2010) is that both models consider stochastic demands. Especially, the second model including correlated stochastic demands addresses an important issue in practice.

In the work of Gaur and Fisher (2004) and Parthanadee and Logendran (2006) it is assumed that the lot size is determined by accumulating the demand until the next scheduled shipment. This is the pull based counterpart to an unlevelled supplier kanban system, in which the demand since the last shipment builds the lot size. Hence, the adaptation of the models and solution heuristics to the kanban case is trivial. In the approach of Gaur and Fisher (2004) the search space is reduced by serving the customers of a cluster either all together or on a point-to-point transport. This is very restrictive, and one might lose optimization potential. Parthanadee and Logendran (2006) propose a PVRP-SC not restricting the tours in a similar way which makes the approach very promising for milk run scheduling.

In the PVRP-SC model of Esparcia-Alcázar et al. (2009) it is assumed that the weekly demand is split into transport lots of equal size also for delivery profiles with varying inter arrival times. This is the push counterpart to a kanban system with heijunka levelling and seems adequate to model volume consistent milk runs. However, the authors only consider cost estimations for the inventory on frequency level and do not explicitly consider the inventory behaviour at the supplier sites and/or at the plant site.

Moreover, apart from "well spaced" weekly patterns none of the proposed models considers further consistency requirements, such as approximately the same arrival times over the week or partial or strict driver consistency. Since these features seem beneficial for a milk run schedule, other VRP models considering consistency and complexity aspects need to be taken into account (see (Francis et al. 2006a), (Groër et al. 2009), (Coelho et al. 2012) or (Spliet 2013)). Please remember that the resulting VRP models can be classified as VRP with synchronization (see (Drexl 2012)).

Transport Concept Assignment To the best of our knowledge, there exists no scientific work addressing both tactical tasks simultaneously:

Selecting the appropriate transport concept and planning the milk run schedules for the assigned suppliers. All approaches dedicated to plan regular schedules described above assume that the suppliers which need to be considered are given. A model addressing both aspects closely resembles the periodic version of the following models: the operational transport concept assignment models of Branke et al. (2007) and Hosseini et al. (2014), which also consider dedicated tours, or the VRP formulations with outsourcing for freight forwarders proposed by Stenger et al. (2013) and Wang (2015).

Frequency Assignment There exist good models in literature for assigning inventory and transport cost optimal cycle times (see Table 5.2) or delivery profiles (see Table 5.3) considering different transport concepts. The models also cover complex tariff structures as they occur, for example, in case of groupage services. Transport concepts requiring the solution of vehicle routing decisions for determining the transport cost, such as milk runs, are not considered. However, the proposed approaches can be used to determine a cost optimal lower bound for the frequency of a milk run:

Let us consider an inbound network providing groupage service with a single tariff table, point-to-point transports and milk runs as transport concepts. In this case, we can assume that a milk run is only established for a group of suppliers if it yields less operational cost compared to an execution using the two alternative transport concepts. That means, the costs for groupage services or direct transports – depending on what transport concept is the cheaper one – can be used as an upper bound for transport cost in the inbound network.

Consequently, the approaches proposed to determine optimal cycle times and delivery profiles considering groupage services or point-to-point transports give a lower bound for the cost optimal frequency for the inbound network. Please note that in case of separate groupage tariffs for the pre- and main-leg the suppliers must be considered one by one in order to receive a true lower bound.

The maximum of this cost optimal lower bound and the lower bound resulting from restricted resources – e.g. storage space or specialized container – can be used as an overall lower bound for the frequency during the Milk Run Design decision process. Moreover, the approaches can be used to evaluate the cost increase for inventory if higher frequencies or delivery profiles are selected.

However, for determining the minimum frequency resulting from restricted resources a model describing the inventory behaviour at the plant and the supplier site is necessary. To the best of our knowledge, also within the "lot sizing community", no explicit inventory model has been proposed considering delivery profiles with varying inter arrival times and equally sized lot sizes as assumed in the milk run scheduling model of Esparcia-Alcázar et al. (2009) or occurring in a heijunka levelled kanban system with weekly shipment patterns.

6.2 Contributions and Outline of Remaining Chapters

For milk runs with a system cycle time of one day, there already exist good approaches. However, in Europe a huge part of the regular supplier relations are served daily or less frequently by groupage services. This is especially the case for suppliers in the automotive industry with considerably smaller supplier networks and shipping volumes. Hence, in the remainder of this thesis we focus on the case of weekly recurring milk run schedules. Due to the planning certainty, the corresponding tours can be easily outsourced to a freight forwarder, and good tariffs can be arranged.

For developing an appropriate Milk Run Design model we focus on the following contributions:

Inventory Model for Varying Inter Arrival Times The goal of *Chapter 7* is to achieve a better understanding of the inventory behaviour at the supplier and the plant sites in case of cyclically repeated transport schedules for both push and pull based systems. The most important contribution of this chapter is the development of a deterministic inventory model for delivery profiles with varying inter arrival times and constant lot sizes reached by heijunka levelling in the pull case and a corresponding order policy for the push case.

Consistency in Milk Run Schedules Regularity and consistency are important features of milk run schedules since they allow for continuous improvement and a smooth flow of parts. In *Chapter 8* we propose different models for determining both weekly milk run schedules with constant inter arrival times – and hence, constant cycle times – and inter arrival time consistent milk run schedules. Additionally, we model driver consistency and partial driver consistency. By applying these models to small instances we determine the cost of different types of consistency.

Transport Concept Assignment According to our experience, point-to-point transports and groupage services are the two transport concepts mainly applied in most automotive companies. Hence, milk runs must compete in terms of operational cost with these two concepts. In *Chapter 9* we extend the aforementioned milk run scheduling models by the decision which suppliers should be assigned to the milk run concept and which should be served by point-to-point transports or groupage service. Please note that, due to the same cost structure, we do not need to differentiate between groupage service with and without delivery profiles from a modelling point of view.

Case Study Bosch Homburg In *Chapter 10* we describe the results of a case study based on historic data of the inbound network of a plant of Bosch in Homburg (Germany). The goal is to evaluate the models which simultaneously determine the transport concept and the milk run schedules in terms of operational cost. We, furthermore, propose an easy to calculate parameter, indicating if the application of a milk run concept as a complement to an existing groupage service network might be beneficial and an indicator measuring the performance of established milk runs in relation to alternative transport concepts.

In order to keep this thesis focussed, some aspects mentioned in the preceding chapters are not considered: Due to the additional complexity of multi echelon routing, we do not consider indirect milk runs. We furthermore assume that groupage services are offered with a single transport tariff table for the pre- and main-leg. This is a common situation in practice. In order to understand the basic behaviour of the models, we do not consider any stochastic effects. However, all aspects should be subject of further research.

7 Deterministic Cyclic Inventory Models for Push and Pull Systems

The goal of this chapter is to achieve a better understanding of the inventory behaviour at the supplier and the customer sites in case of cyclically repeated transport schedules for both push and pull based order systems. In particular, we propose formulas to determine the minimum inventory cost per frequency. Moreover, we derive optimality properties for transport patterns yielding minimum inventory. These patterns and the corresponding cost are used as input for the milk run scheduling models introduced in the succeeding chapters.

We start by introducing finite horizon models – briefly discussed in Section 5.2 – in greater depth since cyclic inventory models are a generalisation. However, these models usually look at push based systems considering only the inventory at the plant. We therefore briefly characterize the differences between pull and push based systems, and between finite horizon and cyclic models. Afterwards we propose the adapted cyclic versions of the finite models and give optimality properties to determine patterns with minimal inventory. We furthermore develop a third model considering levelled transport lot sizes in case of varying inter arrival times occurring, for example, in weekly milk runs.

We consider only deterministic versions since we assume that varying production rates at the supplier are buffered by its final parts stock, while variation in the demand and the transport time is assumed to be buffered through safety inventory at the plant site.

7.1 Finite Horizon Inventory Models with Constant Demand for Push Systems

In Section 5.2.1 we briefly introduced EOQ models for problems with a finite horizon. In Section 7.1.1 we further detail the model assuming that the time in between two shipments is continuously divisible, while the discrete time model is detailed in Section 7.1.2. In the latter case it is assumed that the inter arrival times must be an integer multiple of a base period – for example a day.

7.1.1 A Continuous Time Model

Assumptions As introduced in Section 5.2.1, we assume that the plant has a constant production rate consuming a supply part with a demand rate d during a planning horizon T. We furthermore consider a fixed order (setup) cost c^o and a linear inventory holding cost c^h incurred for every unit on stock per unit time.

Considering the Situation at the Plant Following the introduction of Simchi-Levi et al. (2014a), we only consider the situation at the plant. We assume an inventory policy $\mathcal{P}$ that places at least $f \in \mathbb{N}_{>0}$ orders during the planning horizon $[0, T]$. A look at Figure 7.1 shows that an optimal ordering policy must satisfy the zero-inventory-ordering property. Furthermore, we know that the inventory decreases at a rate of d and reaches zero exactly f times.

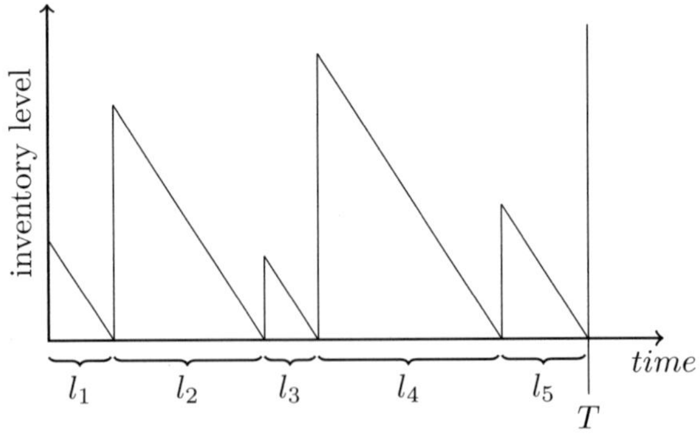

Figure 7.1: Inventory level as a function of time under a policy $\mathcal{P}$ (Source: (Simchi-Levi et al. 2014a))

Let us assume that $l_i > 0$ with $i \in \{1..f\}$ stands for the time in between two shipments and that $\sum_{i=1}^{f} l_i = T$. Then, we can define the inventory level emerging from applying the policy $\mathcal{P}$ by calculating the area of the resulting triangles:

$$\mathcal{I}_{plant} = \sum_{i=1}^{f} \frac{d \cdot l_i \cdot l_i}{2} \tag{7.1}$$

The best times to place f orders can be found by solving

$$min \left\{ \sum_{i=1}^{f} l_i^2 \mid \sum_{i=1}^{f} l_i = T \quad i \in \{1, .., f\} \right\} \tag{7.2}$$

As the optimal solution of this convex optimization problem is $l = \frac{T}{f}$ for all i, an optimal policy $\mathcal{P}$ must fulfil the following property:

Property 7.1.1 *The policy with minimum cost that places f orders during a finite horizon $[0, T]$ is achieved by ordering every $\frac{T}{f}$ time units an order of size $\frac{T}{f} \cdot d$ with the first order in $t = 0$.*

Thus, the cost associated with the optimal policy $\mathcal{P}$ for the planning horizon T and the frequency f is:

$$c^o \cdot f + c^h \sum_{i=1}^{f} \frac{d \cdot l_i \cdot l_i}{2} = c^o \cdot f + c^h \frac{d \cdot T^2}{2 \cdot f} \tag{7.3}$$

7.1.2 A Discrete Time Model

In a finite horizon discrete time model, the times in between shipments can no longer be all of the same size if f is not a divisor of T. For finding the periods and the order quantities the same quadratic program as (7.2) must be solved with the difference that l_i must be integer numbers between 1 and T. Since the zero-inventory-ordering property must be fulfilled as well, it is enough to determine either the inter arrival times or the order sizes.

As introduced shortly in Section 5.2.1, Kovalev and Ng (2008) showed for the finite horizon discrete time EOQ problem considering only the inventory at the plant site that a minimum inventory level can be reached for a given frequency f if one places $r = f \cdot \lceil \frac{T}{f} \rceil$ orders of size $d \cdot \lceil \frac{T}{f} \rceil$ and $f - r$ orders of size $d \cdot \lfloor \frac{T}{f} \rfloor$. That means, the order sizes and inter arrival times take at most two different values (this also corresponds to the independently derived optimality properties of Fleischmann (1999)). The sequence of these f orders has no impact on the cost.

Thus, in the discrete time model the cost associated with an optimal order schedule $\mathcal{P}$ for the planning horizon T and the frequency f is:

$$c^o \cdot f + c^h \cdot \frac{d}{2}\left(r \cdot \left\lfloor \frac{T}{f} \right\rfloor + (f - r) \cdot \left\lceil \frac{T}{f} \right\rceil\right) \tag{7.4}$$

Please note that the solution of the continuous time model results in a schedule referred to in Chapter 5 as constant cycle schedule. The second model results in patterns with differing inter arrival times also referred to as delivery profiles in Chapter 5. However, in this section we stick with the terms continuous and discrete time models in order to be consistent with existing inventory models in literature.

7.2 Inventory Behaviour in Cyclic Pull and Push Systems

In the preceding section we considered a classic model for a push based system. However, the cost optimal inventory behaviour of pull and push based systems in a deterministic finite-horizon setup is very similar: The difference is that in a pull based model the transport lot size is determined by the consumption since the last shipment, while in the push based version it covers the demand until the next shipment. Furthermore, in kanban cycles – besides the inventory at the plant site – always the inventory at the supplier site is taken into account. For push based policies this is not always the case as shown in the introduction of the preceding section.

7.2.1 Pull vs. Push Systems

The simplified examples in Figure 7.2 and Figure 7.3 show the minimum inventory levels over the planning horizon of one week at the supplier site and at the plant site assuming a weekly frequency of 3, a demand and reproduction rate of 12 parts per day, and a transport time of 0 for a pull based kanban system and a push policy. The first figure illustrates that, in case of constant cycle times, the inventory behaviour between kanban and the push based systems is analogue.

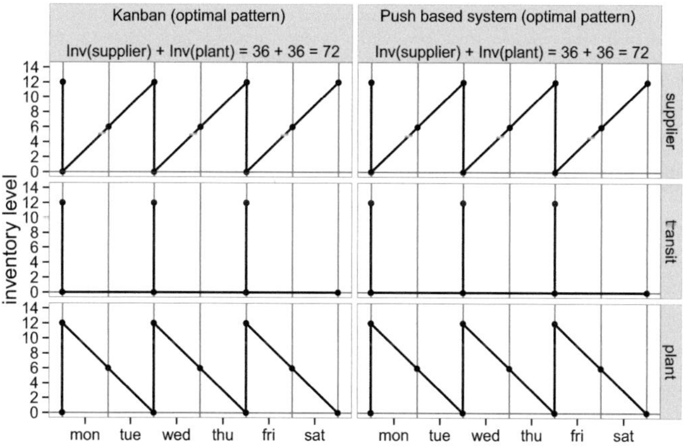

Figure 7.2: Every column shows the inventory level at the supplier (first row), in transit (second row) and at the plant (third row) for a kanban (left column) and a push based (right column) system

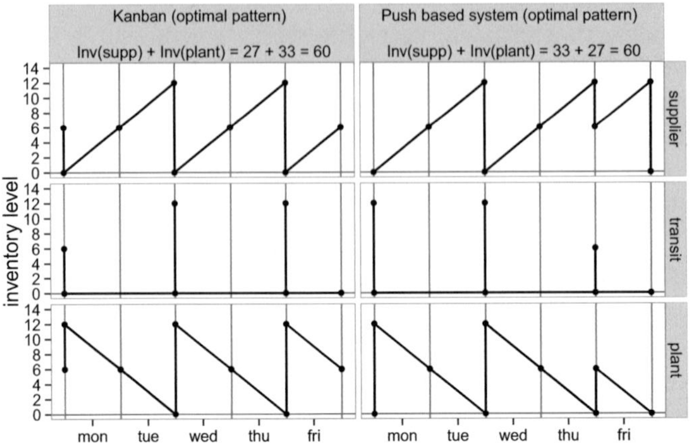

Figure 7.3: Every column shows the inventory level at the supplier (first row), in transit (second row) and at the plant (third row) for a kanban (left column) and a push based (right column) system

In case of varying inter arrival times as shown in Figure 7.3, the same delivery pattern Mon-Wed-Fri leads to differing transport lots. However, a closer look at the inventory levels shows that the inventory represented by the area under the sawtooth pattern is the same: The inventory at the plant in the pull case corresponds to the inventory at the supplier site in the push case and vice versa, due to forming the transport lots either based on the preceding time steps or on the succeeding ones. Hence, in the following we focus on kanban systems and only come back to push based policies in cases when the characteristics are different.

7.2.2 Critical Work in Progress

As introduced in Section 3.3.1, the biggest advantage of all types of kanban systems – beside the ease of implementation – is that the work in progress (WIP) is capped automatically by the number of cards. No unnecessary WIP that does not improve the throughput is produced (see (Hopp and Spearman 2001a)). The so called critical work in progress

(WIP_0) is the minimum number of parts needed to guarantee a supply process without production interruptions at the point of consumption. In case of supplier kanban, it corresponds to the aggregated inventory at the supplier site, the plant site, and in transit.

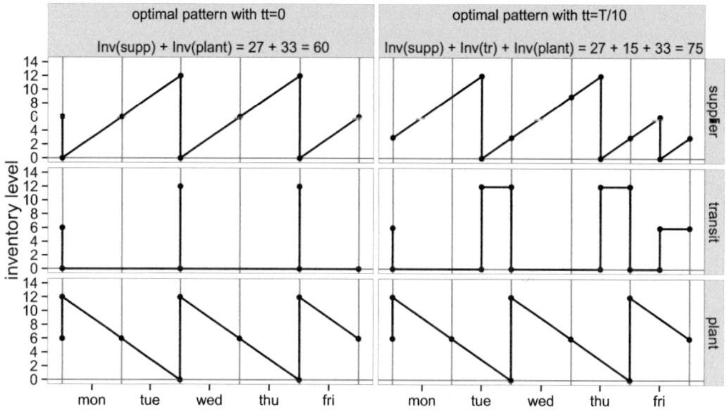

Figure 7.4: Every column shows the inventory level at the supplier (first row), in transit (second row) and at the plant (third row) for a transport time of $tt = 0$ (left column) and transport time of half a day (right column) system

The critical work in progress divided by the kanban unit size M gives the lower bound for the number of kanban cards needed. The number of cards also caps the maximum inventory at one point of the kanban cycle, that means, it caps the maximum inventory at the supplier site and the plant site.

The example of Figure 7.4 corresponds to the example of Figure 7.3. When we sum up the 3 inventory levels of the kanban loop at any point in time, we obtain the WIP_0, and we see that it is constant for all points in time. In the left side example of the figure, the time for the whole shipping process is assumed to be 0 while on the right side it is assumed to be half a day.

7.2.3 Cyclic vs. Finite Time Models

Finite horizon models are introduced for cases in which a part is only needed for a certain time T. At the beginning of such a period, the inventory naturally is zero and at the end, when no more parts of this type are needed, as well. Our case is more general: The transport schedules are repeated after T periods. Hence, the inventory level at the beginning of the planning horizon must correspond to the inventory level at the end of the planning horizon. Initial and final inventory does not need to be 0.

7.3 Modelling Assumptions

Following the assumptions introduced for the finite horizon models in Section 7.1.1 and 7.1.2, we assume for the cyclic models that a plant has a constant consumption rate of $d > 0$ per period. Moreover, we assume that a supplier reproduces the products with the same rate d. The planning horizon of the model is T, and the resulting pattern are cyclically repeated every T periods.

Following the assumption of the continuous time finite horizon models introduced before, we assume that the products are continuously divisible. That means that the possibly non integer lot size of $\frac{T}{f} \cdot d$ (see Section 7.1.1) is valid or that the demand is large enough so that the non integer lot size is a good approximation. This is an assumption not necessarily fulfilled in practice. However, in the following we want to focus on the effects of different ordering policies. Rounding issues, which are in practice reinforced by minimum and/or maximum lot sizes and packaging units, should be considered in a second step and are beyond the scope of this thesis.

For the cyclic models we furthermore assume that eKanban are utilised: The electronic card is detached from a part in the moment when it is withdrawn for production and the re-production signal is transferred to the supplier assuming a transfer time of 0. The production process immediately starts so that the supply part is finished at the same time as a new one is needed at the plant. This can be assumed in an analogue way for a push based policy.

Hence, the models focus on transport induced inventory: The time between shipments is the only reason for holding parts on stock since we do not need any stock for buffering production or communication lead times or safety stock for buffering variability in demand or processes.

We furthermore assume that the inventory in transit is the same for all models and does not influence the decision on a cycle time or pick up schedule. The transport time (tt) simply shifts the inventory level at the supplier by tt as can be seen from the second example of Figure 7.4, and it determines the inventory in transit. If vol_i represents the shipping volume of shipment i, the inventory in transit sums up to $\mathcal{I}_{trans} = vol_1 \cdot tt + ... + vol_f \cdot tt = T \cdot d \cdot tt$, i.e. the inventory in transit does not depend on the frequency or the pattern.

In the following, we furthermore assume that the inventory costs at the supplier and the plant site are the same. Hence, we focus on determining the total minimum inventory, yielding at the same time the total minimum inventory cost. Please note that this relationship is not necessarily valid if the costs are differing. In practice, the assumption of similar inventory cost at the supplier and the plant site is not necessarily fulfilled. However, it seems to be a reasonable simplification since we want to derive simple formulas and optimality properties in order to improve the general understanding of the inventory behaviour in a cyclic setup. Furthermore, in the context of Milk Run Design, inventory cost usually play a negligible role compared to transport cost (see for example (Chuah and Yingling 2005) and (Ohlmann et al. 2008)).

7.4 A Cyclic Continuous Time Inventory Model (CCI)

The extension of the finite model introduced in Section 7.1 to a kanban cycle is straight forward: Besides the inventory at the plant we need to consider the inventory at the supply site. From Figure 7.2 we know that if the orders are equally spaced over the planning horizon the inventory at the supplier behaves symmetrically to the inventory of the plant site and, hence, equally spaced orders of the same size yield minimal inventory cost for the whole system.

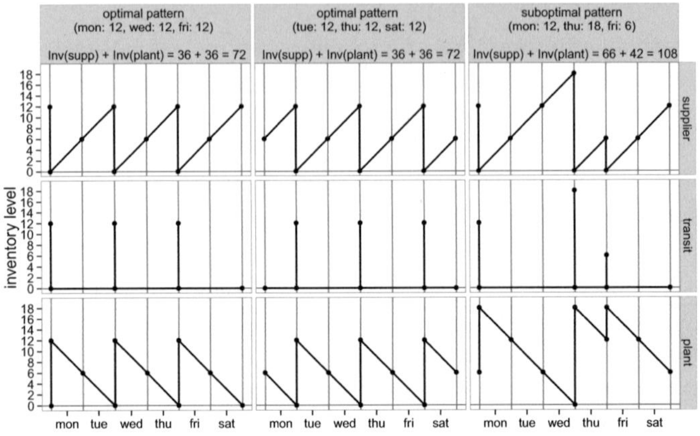

Figure 7.5: Every column shows the inventory level at the supplier (first row), in transit (second row) and at the plant (third row) for different pick up patterns

In case of cyclic inventory models, the first order does not necessarily need to be placed in $t = 0$. It must only be assured that the initial inventory corresponds to the remaining inventory in $t = T$. A look at the example of Figure 7.5 shows that this is also guaranteed for all constant cycle schedules in which the first order is placed between between $t = 0$ and $\frac{T}{f}$. Please note that the first visit must be scheduled before the second. Hence the following property can be formulated:

Property 7.4.1 *For CCI, shipment patterns $\mathcal{P}$ yielding minimum total inventory cost for a frequency of f and a continuous time planning horizon of $[0, T]$ are achieved by distributing the pick ups equally over the planning horizon with constant inter arrival times of size $\frac{T}{f}$. The first shipment can be scheduled within $[0, \frac{T}{f})$.*

Since the inventory at the supplier and the plant are considered for determining the cost, the total cost over the planning horizon associated with an optimal pattern can be given for the planning horizon T and the frequency f as:

$$c^o \cdot f + 2 \cdot c^h \frac{d \cdot T^2}{2 \cdot f} + c^h \cdot T \cdot d \cdot tt = c^o \cdot f + c^h \frac{d \cdot T^2}{f} + c^h \cdot T \cdot d \cdot tt \tag{7.5}$$

Since the critical work in progress for the whole system is constant for every time step t, we can calculate it by dividing the total inventory $\mathcal{I}$ by T resulting in $d \cdot \frac{T}{f}$.

Relationship to the Toyota Formula The Toyota formula to calculate the number of kanban introduced in Chapter 3 assumes constant cycle times and, hence, requires that Property 7.4.1 is fulfilled.

As introduced in Chapter 3, the Toyota formula determines the number of kanban N for a given configuration $a - b - c$ by:

$$N = \left\lceil \frac{a \cdot (c+1)}{b \cdot M} + \frac{S}{M} \right\rceil$$

Assuming a delivery time at the supplier of $c = 0$, a kanban unit $M = 1$ and no safety stock S, we receive the number of kanban cards induced by the transportation process:

$$N^T = d \cdot \frac{a}{b}$$

This corresponds to the $\text{WIP}_0 = d \cdot \frac{T}{f}$ of the optimal order schedule for frequency f. If small deviations from the optimal inter arrival time (or cycle time) $T/f = a/b$ occur, they are included in the delay coefficient c.

7.5 A Cyclic Discrete Time Inventory Model (CDI)

Assuming constant demand, constant cycle times lead to complicated schedules in case of weekly recurring milk runs if f is not a divisor of T. In this section we propose cyclic inventory models assuming discrete time steps or, stated differently, schedules with possibly differing inter arrival times, which are multiples of a base period. In a kanban system and in case of an optimal push based policy – following the zero-inventory-ordering property – not only the inter arrival times vary but also the transport lots.

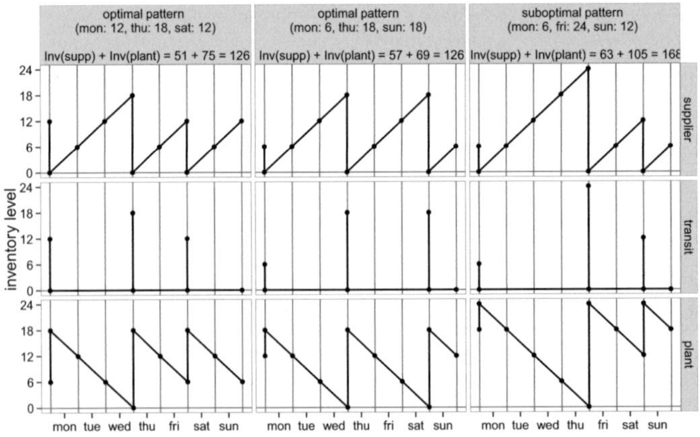

Figure 7.6: Every column shows the inventory level at the supplier (first row), in transit (second row) and at the plant (third row) for different patterns

Thus, in this section the planning horizon T and the inter arrival times l are measured in base periods and must be integer with $l \in \{1, 2, .., T\}$. Let us consider the case $(T \bmod f) \neq 0$, that means the optimality property of constant cycles from the previous section delivers a fractional value: The examples of Figure 7.6 show that the inventory at the supplier site is defined by the size of every single interval since it goes down to zero with every shipment.

However, at the plant site the inventory level is defined by the size of every interval and – at the same time – by the size of the longest interval, as the inventory is built up over the whole system cycle time to provide for this long interval without supply.

For a pattern $\mathcal{P}$ of length T let $l_{max} \in \{1, 2, .., T\}$ denote the longest inter arrival time (measured in base periods). For each $j \in \{1, 2, .., l_{max}\}$ we define $n_j \in \{0, 1, .., f\}$ as the number of inter arrival intervals of length j in pattern $\mathcal{P}$. If no interval of size j is part of the solution, n_j is equal to 0. If these parameters describe a feasible pattern $\mathcal{P}$, the following equations hold:

$$\sum_{j=1}^{l_{max}} n_j = f \tag{7.6}$$

$$\sum_{j=1}^{l_{max}} j \cdot n_j = T \tag{7.7}$$

The inventory of the corresponding pattern can be calculated by applying simple geometric formulas for the supplier and the plant site:

$$\mathcal{I}_{supp} = \sum_{j=1}^{l_{max}} \left(\sum_{t=0}^{j-1} t \cdot d + \frac{1}{2} \cdot d \cdot j \right) \cdot n_j \tag{7.8}$$

$$\mathcal{I}_{plant} = \sum_{j=1}^{l_{max}} \left(\sum_{t=0}^{j-1} (l_{max} - 1 - t) \cdot d + \frac{1}{2} \cdot d \cdot j \right) \cdot n_j \tag{7.9}$$

$$\mathcal{I}_{trans} = T \cdot d \cdot tt \tag{7.10}$$

With the outer summation operator in (7.8) and (7.9) we sum up the inventory over all intervals of the different sizes j with $1 \leq j \leq l_{max}$. We multiply this area with the number of times n_j that an interval of size j exists in $\mathcal{P}$. Please note that n_j for the interval size $j = l_{max}$ is always at least 1 since the interval of maximal size must exist per definition.

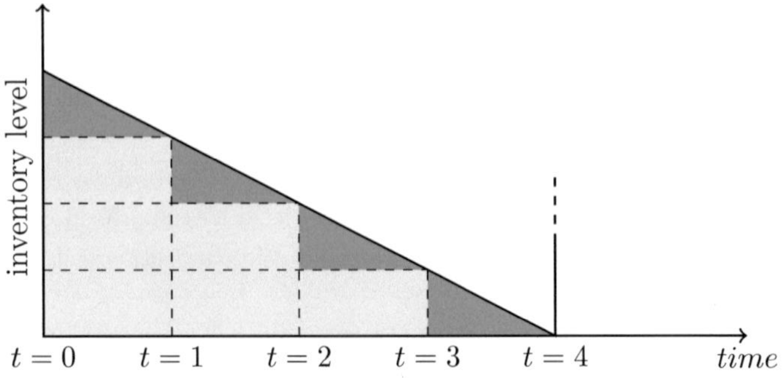

Figure 7.7: Example illustrating the different components of the formulas (7.8) and (7.9) calculating $\mathcal{I}$ with the squares (light grey) resulting from the first part of the summation operator and the triangles (dark grey) from the second

From Figure 7.7 one can see that the first part of the inner summation operator assures that the area of the squares is considered. This is the part of the inventory only built up to provide for periods without shipment. The second summand gives the area of the triangles. These triangles occur in every base period independently of the frequency f since the inventory is built up and reduced linearly over the course of the base period.

If we aggregate the inventories of the whole kanban cycle, we get:

$$\mathcal{I}_{total} = \sum_{j=1}^{l_{max}} \left((\sum_{t=0}^{j-1}(t + l_{max} - 1 - t) \cdot d \cdot n_j) + d \cdot j \cdot n_j \right) + T \cdot d \cdot tt \tag{7.11}$$

$$= \sum_{j=1}^{l_{max}} ((l_{max} - 1) \cdot d \cdot j \cdot n_i + d \cdot i \cdot n_i) + T \cdot d \cdot tt \tag{7.12}$$

$$= \sum_{j=1}^{l_{max}} (l_{max} \cdot d \cdot j \cdot n_j) + T \cdot d \cdot tt \tag{7.13}$$

If we substitute $\sum_{j=1}^{l_{max}}(j \cdot n_j)$ by T, formula (7.13) simplifies to:

$$\mathcal{I}_{total} = l_{max} \cdot T \cdot d + T \cdot d \cdot tt \tag{7.14}$$

This shows, as expected from Chapter 3, that the total inventory in a kanban cycle induced by the transport schedule depends only on the longest interval between two consecutive shipments, and patterns which yield a minimum total inventory minimize the maximum length of all intervals.

Since we require that the inter arrival times, which we measure in base periods, must be integer, the maximum period l_{max} between two shipments must be at least $\lceil \frac{T}{f} \rceil$ because $\lfloor \frac{T}{f} \rfloor \cdot f < T$ if $(T \bmod f) \neq 0$ holds. Consequently, given f and T, the minimum total inventory is determined by:

$$\mathcal{I}_{total} = \left\lceil \frac{T}{f} \right\rceil \cdot T \cdot d + T \cdot d \cdot tt \tag{7.15}$$

Whenever $(T \bmod f) = 0$ holds, the resulting inventory corresponds to the optimal case of constant inter arrival (cycle) times. Hence, the formula is also applicable for the case in which f is a divisor of T.

Since the critical work in progress for the whole system is constant for every time step t, we can calculate it by dividing $\mathcal{I}_{total}$ by T resulting in $\text{WIP}_0 = d \cdot (l_{max} + tt)$. It gives, at the same time, the maximal storage place needed at the supplier and the plant site required in the moment when all inventory is either at the supplier or the plant site.

Minimizing Total Inventory With these insights, we can derive a constraint satisfaction problem (CSP) defining the set of schedules yielding minimum inventory levels for a given frequency f and a planning period of T. If s_i stands for the period in which the i-th shipment takes place, the CSP is defined by the following set of constraints:

$$s_{i+1} - s_i \leq \left\lceil \frac{T}{f} \right\rceil \qquad i = 1, .., f-1 \tag{7.16}$$

$$s_1 + T - s_f \leq \left\lceil \frac{T}{f} \right\rceil \tag{7.17}$$

$$s_i \leq s_{i+1} \qquad i = 0, .., f-1 \tag{7.18}$$

$$s_i \in \{0, T-1\} \qquad i = 0, .., f \tag{7.19}$$

With constraints (7.16) we assure for shipment 1 to f that the time between two consecutive shipments is not greater than $\lceil \frac{T}{f} \rceil$. Since the schedules are repeated, the next constraint set imposes this requirement for the last and the first shipment. Constraints (7.18) break symmetry, while the last line sets the domains of variables s_i. If $(T \bmod f) = 0$ holds, all inter arrival times are just of length $\frac{T}{f}$. Hence, an optimal schedule $\mathcal{P}$ must have the following property:

Property 7.5.1 *For CDI, a pattern $\mathcal{P}$ for a time discrete planning horizon $\{0, 1, .., T\}$ with inter arrival times being integer (measured in base periods) yields minimum total inventory for a frequency f if it is assured that the maximum inter arrival time is smaller or equal to $\lceil \frac{T}{f} \rceil$. All patterns fulfilling this property are described by constraint sets (7.16)-(7.19).*

Minimizing Total Inventory and Inventory at the Plant Site An important insight of the decomposed inventory formulas (7.8) and (7.9) is that – in cases in which $(T \bmod f) \neq 0$ holds – the average inventory at the plant site is always higher than at the supply site. Remember that, in case of a push based policy, it is the other way around.

Furthermore, different patterns yielding minimum total inventory result in different inventory shares between the supplier and the plant: In case of the inventory pattern in the first column of Figure 7.6, around 40% of the weekly inventory is allocated at the supplier and 60% at the plant, while in case of the second column 45% are allocated at the supplier and only 55% at the plant.

From a supply chain management point of view, it is beneficial to store parts as close as possible to the point of consumption in order to reduce the risk of backlogs. Consequently, the pattern of the first example of Figure 7.6 can be considered as better. In general, that means a pattern

$\mathcal{P}$ is preferable if it minimizes the total inventory and, at the same time, the inventory at the supplier site.

As mentioned in Section 7.2, the inventory at the supply site in a kanban case corresponds to the situation at the plant site under a push based policy. According to the discrete time model of Kovalev and Ng (2008), a schedule with only two interval sizes $\lceil \frac{T}{f} \rceil$ and $\lfloor \frac{T}{f} \rfloor$ yields minimum inventories in this case. This further restricts the number of inventory optimal patterns described by the constraint sets (7.16) to (7.19).

Consequently, a minimum total inventory pattern $\mathcal{P}$ which minimises also the supplier inventory must fulfill in addition to the constraint sets (7.16) to (7.19) the following restrictions:

$$s_{i+1} - s_i \geq \left\lfloor \frac{T}{f} \right\rfloor \qquad i = 1, .., f-1 \tag{7.20}$$

$$s_1 + T - s_f \geq \left\lfloor \frac{T}{f} \right\rfloor \tag{7.21}$$

Property 7.5.1 is extended accordingly for this case.

Since we assumed that inventory costs at the plant site and the supply site are the same, we can give the costs for a frequency f and a planning period T for both variants as:

$$c^o \cdot f + c^h \cdot \left\lceil \frac{T}{f} \right\rceil \cdot T \cdot d \cdot + c^h \cdot T \cdot d \cdot tt \tag{7.22}$$

7.6 A Cyclic Discrete Time Inventory Model with Levelled Transport Lots (CDI-LT)

The strategy for the CDI model of the preceding section leads to varying lot sizes. However, since consistent shipping volumes might be wanted to achieve a levelled use of transport resources, we introduce a third model: Heijunka levelling is applied in a cyclic discrete time inventory system in order to achieve equally sized transport lots. To the best of our knowledge, there exists no finite or cyclic inventory model considering such an ordering strategy resulting in levelled lot sizes for all patterns. The same model can be easily applied to a push based order policy.

Figure 7.8 shows the differing inventory behaviours of a heijunka levelled kanban loop compared to the case of unlevelled cycles considered in the preceding section. It presents a case in which f is not a divisor of T. One aspect immediately attracts attention: If the shipping lots are of equal size, the number of times the inventory goes down to zero at both sites is smaller. This is true for all patterns in which f is not a divisor of T.

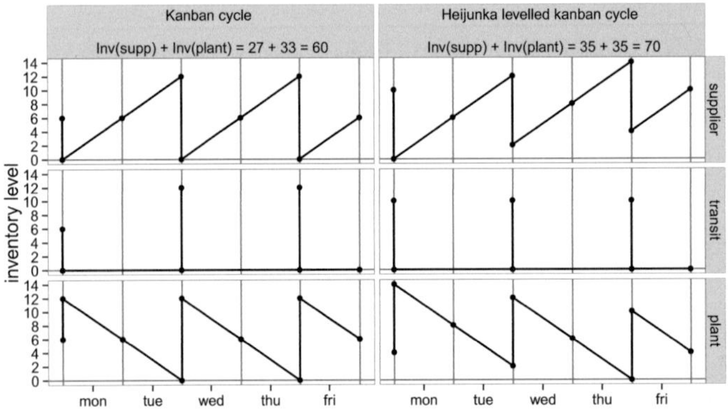

Figure 7.8: Every column shows the inventory level at the supplier (first row), in transit (second row) and at the plant (third row) for a kanban controlled cycle and a heijunka levelled kanban cycle

However, as in the preceding case, the maximum time between two consecutive shipments influences to a great extent the minimum inventory needed and, therefore, should not be longer than $\lceil \frac{T}{f} \rceil$. Since the inventory does not go down to zero as often as in the unlevelled case, the number of long intervals and the sequence of long and shorter intervals play a role.

This becomes clear with the following considerations: Within the interval of size $\lceil \frac{T}{f} \rceil$ the production aggregates to $\lceil \frac{T}{f} \rceil \cdot d$, while at the end only $\frac{T}{f} \cdot d$ units have been shipped. That means, at the end the excess production

$\Delta^+ = (\lceil \frac{T}{f} \rceil - \frac{T}{f}) \cdot d$ is used to build up stock in order to be prepared for a short interval in which the production is below the shipping volume and at least $\Delta^- = (\lfloor \frac{T}{f} \rfloor - \frac{T}{f}) \cdot d$ products from stock are needed at the end of the period for the shipment (see also the second column of Figure 7.9). During the whole planning horizon T, not more than the total demand is delivered, and the sum over all Δ^- and Δ^+ must equal to zero.

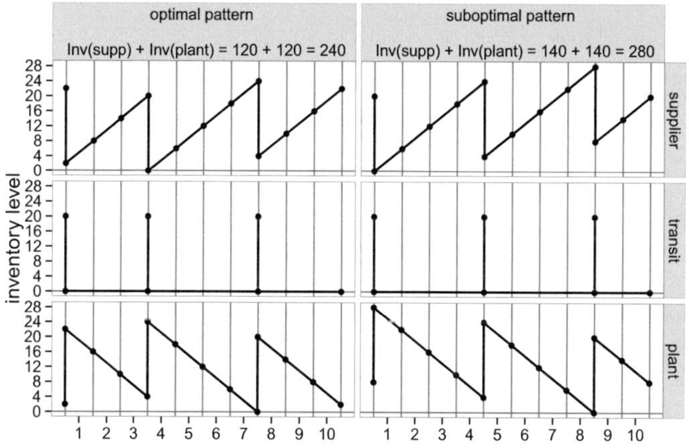

Figure 7.9: Every column shows the inventory level at the supplier (first row), in transit (second row) and at the plant (third row) for different patterns

For the plant these considerations are valid in an analogue way: The plant receives more parts than necessary for the shorter intervals and builds up stock for consumption during the long interval. This means that the inventory in a heijunka levelled system is always equally distributed between the supplier and the plant site. Therefore, in the following we focus on the supplier site without loss of generality.

The volume produced during an interval should be as close as possible to the shipping volume: Ideally it should be either slightly above – with $\lceil \frac{T}{f} \rceil \cdot d$ – or slightly below – with $\lfloor \frac{T}{f} \rfloor \cdot d$. This is fulfilled in the example of the first column of Figure 7.9. Moreover, the number of long intervals, in which overproduction occurs, must be minimal.

We will now show that, for given values of T and f, it is always possible to find a valid pattern with $nl \in \mathbb{N}$ intervals of size $\lceil \frac{T}{f} \rceil$ and $ns \in \mathbb{N}$ intervals of size $\lfloor \frac{T}{f} \rfloor$ with a scheduled shipment at the end of every interval and with the sum over all Δ^+ and Δ^- of the pattern being equal to 0. ns and nl must sum up to f since a shipment is scheduled at the end of each interval.

Given T and f, nl and ns, which satisfy equations (7.23) and (7.24), must be non-negative integers.

$$\Delta^+ \cdot nl + \Delta^- \cdot ns = 0 \tag{7.23}$$

$$ns + nl = f \tag{7.24}$$

Considering the formulas for Δ^+ and Δ^- and inserting $ns = f - nl$ in the first line gives us:

$$\begin{aligned}
\left(\left\lceil \frac{T}{f} \right\rceil - \frac{T}{f}\right) \cdot d \cdot nl + \left(\left\lfloor \frac{T}{f} \right\rfloor - \frac{T}{f}\right) \cdot d \cdot (f - nl) = 0 \\
\overset{d \neq 0}{\Leftrightarrow} \left\lceil \frac{T}{f} \right\rceil \cdot nl - \frac{T}{f} \cdot nl + \left\lfloor \frac{T}{f} \right\rfloor \cdot f - \frac{T}{f} \cdot f - \left\lfloor \frac{T}{f} \right\rfloor \cdot nl + \frac{T}{f} \cdot nl = 0 \\
\Leftrightarrow (\left\lceil \frac{T}{f} \right\rceil - \left\lfloor \frac{T}{f} \right\rfloor) \cdot nl + \left\lfloor \frac{T}{f} \right\rfloor \cdot f - T = 0 \\
\Leftrightarrow nl + \left\lfloor \frac{T}{f} \right\rfloor \cdot f - T = 0 \\
\Leftrightarrow nl = T - \left\lfloor \frac{T}{f} \right\rfloor \cdot f \\
\Rightarrow nl = T \bmod f
\end{aligned}$$

The remainder of an Euclidean division is always integer and smaller than its divisor f.

$$ns = f - nl$$

$$\Leftrightarrow ns = \left\lceil \frac{T}{f} \right\rceil \cdot f - T$$

Since f and nl are integer values, ns is also integer. Thus, it is always possible to compose a feasible pattern of nl long intervals of length $\lceil \frac{T}{f} \rceil$ and ns smaller intervals of size $\lfloor \frac{T}{f} \rfloor$.

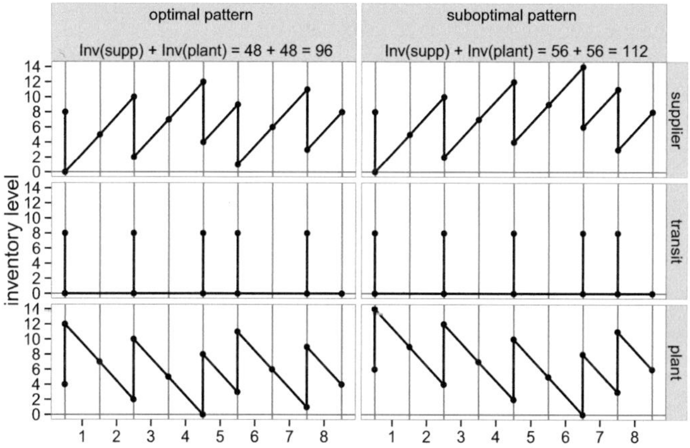

Figure 7.10: Every column shows the inventory level at the supplier (first row), in transit (second row) and at the plant (third row) for different patterns

For the parameters $T = 8$ and $f = 5$, our formula yields three long intervals and two short intervals. Having a look at the schedules of Figure 7.10, our observation from above is further supported: The way how short and long intervals are ordered influences the inventory levels as well. In a long interval, production exceeds consumption, whereas in a short interval consumption exceeds production. Therefore, it is beneïňĄcial if a short interval follows directly after a long one because the excess stock of the long period is immediately consumed in the short one.

In general, we can say that long and short intervals should alternate in a way that the overproduction of a long interval is consumed in subsequent short intervals. Or stated differently: Considering that nl and ns do not have to be equal, we strive for an even distribution of the long intervals over a cyclic pattern. The observation from above can be summarized as:

Property 7.6.1 *Shipment patterns yielding minimum total inventory cost for a frequency f, for a time discrete planning horizon $\{0, 1, .., T\}$ and with equally sized transport lots of size $T \cdot d/f$ have $nl = T \bmod f$ long intervals of size $\lceil \frac{T}{f} \rceil$ and $ns = f - nl$ short intervals of size $\lfloor \frac{T}{f} \rfloor$. The long intervals must be as evenly distributed as possible.*

Since it is difficult to describe the property "as evenly distributed as possible" generally – i.e. for different values of T and f – by constraints as for the properties in the preceding section, we give in the following an efficient algorithm to determine the patterns described by Property 7.6.1:

Algorithm to Determine Inventory Minimal Pattern The task of this algorithm is to determine f shipping periods within the planning horizon $\{0, 1, .., T\}$ in a way that the shipments are evenly distributed building either a long interval or a short one. Summarizing the formulas above to calculate the number of long and short intervals as well as the length of the long (ll) and the length of the short interval (ls) we get:

$$nl = T \bmod f \qquad ll = \left\lceil \frac{T}{f} \right\rceil \qquad l = [b_i | i \in \{1, 2, .., ll\}]$$

$$\text{with} \quad b_i = \begin{cases} 1 & \forall \quad i = ll \\ 0 & \text{else} \end{cases}$$

$$ns = f - nl \qquad ls = \left\lfloor \frac{T}{f} \right\rfloor \qquad s = [b_i | i \in \{1, 2, .., ls\}]$$

$$\text{with} \quad b_i = \begin{cases} 1 & \forall \quad i = ls \\ 0 & \text{else} \end{cases}$$

The arrays l and s represent long and short intervals respectively each with a shipment scheduled at the end: 1 stands for a period with a scheduled shipment and 0 for one without, that means all arrays are of the form [1], [0, 1], or [0, .., 0, 1]. As we consider the case (T mod f) $\neq 0$, the number of long intervals and the number of short intervals is greater than 0.

That means, instead of placing f shipments within the planning horizon, the task can be described as distributing the nl long intervals l and the ns short intervals s in a way that they are alternating as much as possible.

This corresponds to the problem of the first step: We look for a position for the nl long intervals l in an array of size f in a way that l is equally distributed. Therefore, l along with one or more s builds a long interval l', and l along with none, one or more intervals builds a short interval s' – obviously containing fewer short intervals than l'. If we substitute T by f and f by nl in the formulas above, we can determine the number (nl') and the length (ll') of long intervals and the number (ns') and length (ls') of short intervals by:

$$nl' = f \bmod nl \qquad ll' = \left\lceil \frac{f}{nl} \right\rceil \qquad l' = [b_i | i \in \{1, 2, .., ll\}]$$

$$\text{with} \quad b_i = \begin{cases} l & \forall \ i = 0 \\ s & \text{else} \end{cases}$$

$$ns' = f - nl' \qquad ls' = \left\lfloor \frac{f}{nl} \right\rfloor \qquad s' = [b_i | i \in \{1, 2, .., ls\}]$$

$$\text{with} \quad b_i = \begin{cases} l & \forall \ i = 0 \\ s & \text{else} \end{cases}$$

The interval arrays itself are of size ls' and ll' respectively and are filled by one long interval l at the beginning followed by the short intervals s that consume the excess production of the long interval.

If the number of the long intervals nl' is still greater than zero, we repeat this procedure by calculating $nl'' = nl' \bmod nl$, $ll'' = \lceil nl''/nl' \rceil$ and so forth. If $nl' = 0$, we are done since we have only ns' short intervals left consisting of the long and short intervals of the preceding steps. If we concatenate the resulting short interval ns' times, the pattern has size T.

Input: $a > 0 \wedge b > 0$
while $b \neq 0$ **do**
    $temp = a \bmod b$;
    $a = b$;
    $b = temp$;
end
return a;

Algorithm 1: Euclidean Algorithm to calculate $gcd(T, f)$

This recursive approach is an extension of the Euclidean Algorithm for calculating the greatest common divisor *gcd* presented – in its not recursive version – in Algorithm 1: In our approach, T corresponds to variable a and f to b. The length of the long interval nl corresponds to the *temp* variable of the *gcd* algorithm. Obviously, *temp* and, hence, f (see last line of *gcd*) become 0 in the last step of the main while loop, otherwise the *gcd* procedure would not terminate.

$p = []$ ▷ initial pattern of form $[0, .., 0, 1]$;
for $i = 0 \rightarrow length - 1$ **do**
    $p.pushback(0)$;
end
$p.pushback(1)$;
return p;

Function $getInitPat$(length)

The whole algorithm to calculate a pattern in which the long periods are "as evenly distributed as possible" is given in Algorithm 2. Please note that lines (4), (16), and (17) and the while loop correspond exactly to Algorithm 1.

```
p = []                                ▷ pattern of form [l, s, .., s];
p.pushback(long);
for i = 0 → length − 1 do
|   p.pushback(short);
end
return p;
```

Function *getPat*(length, long, short)

```
Input: T > 0 ∧ 0 < f ≤ T
Output: pat
c = 0;
while f ≠ 0 do
    c = c + 1;
    nl = T mod f;
    ns = f − nl;
    ll = ⌈T/f⌉;
    ls = ⌊T/f⌋;
    if c = 1 then
        l = getInitPat(ll);
        s = getInitPat(ls);
    else
        newl = getPat(ll, l, s);
        s = getPat(ls, l, s);
        l = newl;
    end
    T = f;
    f = nl;
end
pat = [];
for i = 0 → ns do pat.pushback(s);
return pat;
```

Algorithm 2: Define heijunka schedule $dhs(T, f)$

With the first step into the while loop ($c = 1$) the patterns l and s are initialized by zeros and ones for periods with and without shipment using Function *getInitPat*. Later the patterns of the preceding step are inserted into the patterns of the current step by calling Function *getPat*.

The greatest common divisor corresponds to the number of the short intervals in the last step – in the following defined as $\bar{ns}$ and $\bar{s}$. That means, if the greatest common divisor is 1, only one short interval $\bar{s}$ builds the whole pattern of size T. If $gcd(T, f) > 1$, the interval $\bar{s}$ is concatenated $gcd(T, f)$ times.

Consequently, the algorithm divides the problem of positioning the long intervals in $gcd(T, f)$ identical subproblems with the problem parameters $T' = T/gcd(T, f)$ and $f' = f/gcd(T, f)$. The excess volume of the long interval is guaranteed to be consumed during the short periods of this subproblem, otherwise the sum of the saldo of the identical subproblems would not be zero. Hence, we can conclude that the number of times the inventory drops to zero corresponds to the greatest common divisor of T and f.

The length of $\bar{s}$ gives the number of patterns fulfilling the requirement of equally distributed long periods: We can shift the start point of $\bar{s}$ to each of its own positions without repeating the sequence of the whole pattern. Trying to exchange a 0 and a 1 in order to create a new inventory optimal schedule, we either receive a schedule which can be created by shifting the starting position of the pattern or a pattern which is not optimal with regard to the total inventory level. This becomes obvious with the examples of Table 7.1.

In summary, it can be said that the described algorithm along with the shifting procedure efficiently determines all inventory optimal schedules for combination of a system cycle time and a frequency.

Please note that the following special case of the problem setting described in (Fleischmann 1999) (see Section 5.2.1) corresponds to the CDI-LT model: The cycle time is fixed to T and the vehicle capacity is set to the levelled lot size $\frac{T}{f} \cdot d$. For this case, Fleischmann (1999) shows, that a pattern $\mathcal{P}$ with the following shipping periods s_i is optimal in terms of inventory:

$$s_i = \left\lceil \frac{i \cdot T}{f} \right\rceil \qquad i = 1, .., f \tag{7.25}$$

The resulting pattern fulfils the optimality Property 7.6.1 for the CDI-LT, as well as, the independently derived optimality properties of (Kovalev and Ng 2008) being the base for Property 7.5.1.

Numerical Examples The numerical example in Table 7.2 shows all steps with intermediate results of *dhs* for $T = 8$ and $f = 5$ in order to illustrate the way how the algorithm works. In this case, the first steps yield that the optimal pattern has 3 long intervals of size 2 and 2 short ones of size 1. In this case, an alternating sequence could be $[l, s, l, s, l]$ or $[l, s, l, l, s]$. That means, in this step we look for 3 positions for the long interval in an array of size f. This results in step 2, after which we need to position 2 long intervals in an array of size 3. In step 4, the number of long intervals is 0, and we are left with one small interval of size 2. Inserting (recursively) all interval arrays of the preceding steps, we end up with the interval $s''' = [0, 1, 1, 0, 1, 0, 1, 1]$, which is of length $T = 8$ – as needed. In this pattern, the long and short intervals are distributed "as evenly as possible". The recursive insertion steps starting at the last step are shown in Table 7.3.

As the single interval s''' can start in each period without changing the sequence, we can create $T = 8$ different schedules from this pattern.

Table 7.1: Examples for shifting the starting point of the pattern marked by a frame to each position of the first short period having length 5 in the left table and length 2 in the right one

0	1	2	3	4
[0]	1	1	0	1
1	[0]	1	1	0
0	1	[0]	1	1
1	0	1	[0]	1
1	1	0	1	[0]

0	1	2	3	4	5
[0]	1	0	1	0	1
1	[0]	1	0	1	0

Table 7.2: Results for $dhs(8,5)$ by steps

Number of intervals	Length of intervals	Interval arrays
STEP 1		
$nl = 8 \bmod 5 = 3$	$ll = \lceil \frac{8}{5} \rceil = 2$	$l = [0,1]$
$ns = 5 - 3 = 2$	$ls = \lfloor \frac{8}{5} \rfloor = 1$	$s = [1]$
STEP 2		
$nl' = 5 \bmod 3 = 2$	$ll' = \lceil \frac{5}{3} \rceil = 2$	$l' = [l,s] = [0,1,1]$
$ns' = 3 - 2 = 1$	$ls' = \lfloor \frac{5}{3} \rfloor = 1$	$s' = [l] = [0,1]$
STEP 3		
$nl'' = 3 \bmod 2 = 1$	$ll'' = \lceil \frac{3}{2} \rceil = 2$	$l'' = [l',s'] = [0,1,1,0,1]$
$ns'' = 2 - 1 = 1$	$ls'' = \lfloor \frac{3}{2} \rfloor = 1$	$s'' = [l'] = [0,1,1]$
STEP 4		
$nl''' = 2 \bmod 1 = 0$	$ll''' = \lceil \frac{2}{1} \rceil = 2$	
$ns''' = 1 - 0 = 1$	$ls''' = \lfloor \frac{2}{1} \rfloor = 2$	$s''' = [l'',s''] = [0,1,1,0,1,0,1,1]$

In Figure 7.11, the inventory curves for one of the 8 optimal schedules for the example is shown in the first column. As $gcd(8,5) = 1$ holds, the inventory level goes down to zero only once at different moments at the supplier and the plant. Whereas, in the second example with a $gcd(8,6) = 2$ the inventory level drops down to zero twice.

Determining Inventory Levels for Optimal Heijunka Patterns In order to derive cost and a lower bound for the number of kanban needed in the heijunka case, we need to determine the total inventory $\mathcal{I}_{tot}$ for the patterns derived by the *dhs* algorithm. In the example of Figure 7.12, the inventory built up for buffering the time until the next shipment corresponds to the shaded rectangles. These rectangles have a width of one period, and the height is always a multiple of $\frac{d}{f}$. If $gcd(T, f) = 1$ holds, the multipliers can be given by $\{0, 1, .., T-1\}$.

If $gcd(T, f) > 1$, as shown before, we can decompose the problem in $gcd(T, f)$ smaller problems with parameters $T' = T/\gcd(T, f)$ and $f' = f/gcd(T, f)$.

Table 7.3: Recursive insertion of the pattern

<table>
<tr><th>Periods</th><th>0</th><th>1</th><th>2</th><th>3</th><th>4</th><th>5</th><th>6</th><th>7</th></tr>
<tr><td>step 5</td><td colspan="8">s'''</td></tr>
<tr><td>step 4</td><td colspan="5">l''</td><td colspan="3">s''</td></tr>
<tr><td>step 3</td><td colspan="3">l'</td><td colspan="2">s'</td><td colspan="3">l'</td></tr>
<tr><td>step 2</td><td colspan="2">l</td><td>s</td><td colspan="2">l</td><td colspan="2">l</td><td>s</td></tr>
<tr><td>step 1</td><td>0</td><td>1</td><td>1</td><td>0</td><td>1</td><td>0</td><td>1</td><td>1</td></tr>
</table>

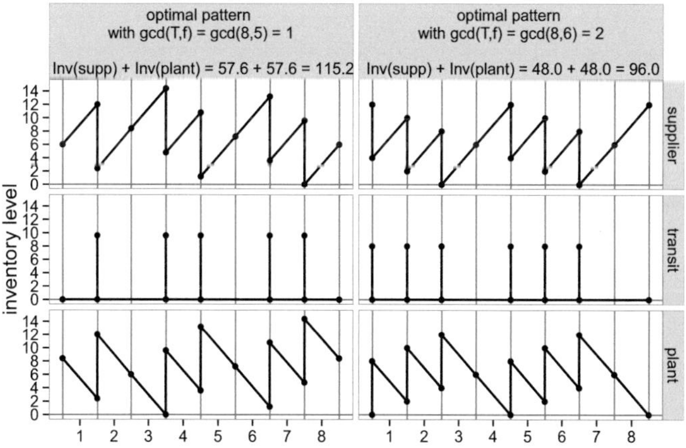

Figure 7.11: Every column shows the inventory level at the supplier (first row), in transit (second row) and at the plant (third row) for different patterns

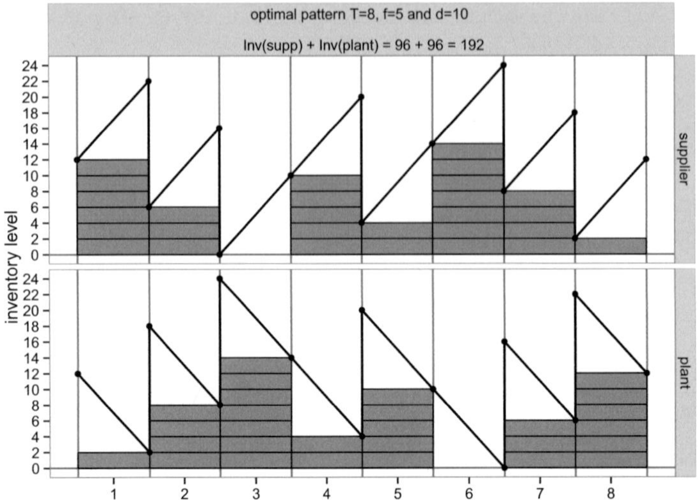

Figure 7.12: Inventory levels at supplier and plant site with the part of the inventory shaded which is built up as there is not a shipment every period

Hence, we introduce the parameters nsc and lsc:

$$nsc = gcd(T, f) \qquad \text{number of repetitions of sub cycles}$$

$$lsc = \frac{T}{nsc} \qquad \text{length of sub cycles}$$

Based upon these parameters we can give the inventory $\mathcal{I}$ using simple geometric formulas for calculating the areas of rectangles and triangles:

$$\mathcal{I}_{supp} = \frac{d}{\frac{f}{nsc}} \cdot nsc \sum_{t=0}^{(lsc-1)} t + \frac{1}{2} \cdot T \cdot d \tag{7.26}$$

$$\mathcal{I}_{plant} = \frac{d}{\frac{f}{nsc}} \cdot nsc \sum_{t=0}^{(lsc-1)} t + \frac{1}{2} \cdot T \cdot d \tag{7.27}$$

$$\mathcal{I}_{trans} = T \cdot d \cdot tt \tag{7.28}$$

As the inventory levels are symmetric, the formula for $\mathcal{I}_{supp}$ corresponds to $\mathcal{I}_{plant}$. The first summand represents the shaded areas in the example, whereas the second summand represents the area of the triangles in the figure. Summing up and simplifying this formula, the total inventory can be given as:

$$\mathcal{I}_{total} = 2 \cdot (\frac{d}{\frac{f}{nsc}} \cdot nsc \sum_{t=0}^{(lsc-1)} t + \frac{1}{2} \cdot T \cdot d) + T \cdot d \cdot tt \tag{7.29}$$

$$= 2 \cdot (\frac{nsc^2 \cdot d}{f} \cdot \frac{(lsc-1) \cdot lsc}{2} + \frac{1}{2} \cdot T \cdot d) + T \cdot d \cdot tt \tag{7.30}$$

$$= (\frac{T - nsc + f}{f}) \cdot T \cdot d + T \cdot d \cdot tt \tag{7.31}$$

$$= (\frac{T - gcd(T,f) + f}{f}) \cdot T \cdot d + T \cdot d \cdot tt \tag{7.32}$$

For cases in which $(T \bmod f) = 0$ and, hence, $gcd(T,f) = f$ holds, the formula results to:

$$\mathcal{I}_{total} = \frac{T}{f} \cdot T \cdot d + T \cdot d \cdot tt \tag{7.33}$$

This corresponds, as expected, to the optimal inventory of constant cycle systems.

Again, the critical work in progress WIP_0 for the whole system can be given by dividing $\mathcal{I}_{tot}$ by T. It is, at the same time, the maximum inventory level at both the supplier and the plant.

The total cost for an inventory optimal schedule with frequency f and for a planning period T can be given by:

$$c^o \cdot f + c^h \cdot \frac{T - gcd(T, f) + f}{f} \cdot T \cdot d + c^h \cdot T \cdot d \cdot tt$$

7.7 Comparison of Cyclic Inventory Models

In summary, we introduced three different cyclic inventory models for the deterministic case of kanban cycles, which are also applicable to push based systems: CCI, CDI, and CDI-LT. In the following we discuss the three underlying strategies differentiated by the categories inventory behaviour, volume consistency, and schedule compatibility. For a better overview, all characteristics are listed in Table 7.6 at the end of this section.

7.7.1 Inventory Behaviour

In the first line of Table 7.6, the total resulting inventories – assuming a transport time of $tt = 0$ – are given. The formulas show that with regard to the inventory levels the constant cycle strategy dominates CDI, which itself dominates the CDI-TL.

To illustrate the difference for concrete planning horizons, in Table 7.4 we give the total inventory depending on the demand rate d for the different inventory models and for three system cycle times T and the corresponding frequencies f. In Figure 7.13 and Figure 7.14, the same information is given in plots assuming a demand rate of $d = 1$.

These results show that the maximum difference of inventory levels between CCI and CDI/CDI-LT is always reached for the frequency of size $f = T - 1$ and that it increases with the planning horizon: For $T = 5$, the increase is around 59%, for $T = 8$ it is around 76%, and for $T = 16$ it is 87%.

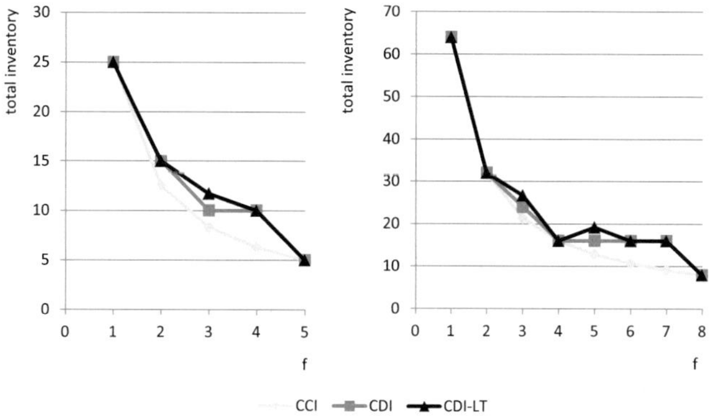

Figure 7.13: Total inventory as a function of the frequency for a planning horizon of $T = 5$ on the left and $T = 8$ on the right side assuming a demand of $d = 1$

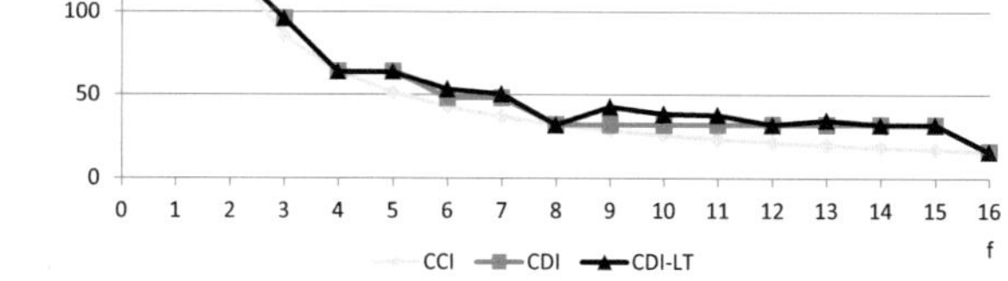

Figure 7.14: Total inventory as a function of the frequency for a planning horizon of $T = 16$ assuming a demand of $d = 1$

Table 7.4: Total inventory $\mathcal{I}_{total}$ depending on the demand rate d for different system cycle times T and frequencies f assuming a transport time of $tt = 0$ differentiated by model type

	T=5			T=8			T=16		
f	CCI	CDI	CDI-LT	CCI	CDI	CDI-LT	CCI	CDI	CDI-LT
1	25 d	25 d	25 d	64 d	64 d	64 d	256 d	256 d	256 d
2	12.5 d	15 d	15 d	32 d	32 d	32 d	128 d	128 d	128 d
3	8.3 d	10 d	11.7 d	21.3 d	24 d	26.7 d	85.3 d	96 d	96 d
4	6.3 d	10 d	10 d	16 d	16 d	16 d	64 d	64 d	64 d
5	5 d	5 d	5 d	12.8 d	16 d	19.2 d	51.2 d	64 d	64 d
6				10.7 d	16 d	16 d	42.7 d	48 d	53.3 d
7				9.1 d	16 d	16 d	36.6 d	48 d	50.3 d
8				8 d	8 d	8 d	32 d	32 d	32 d
9							28.4 d	32 d	42.7 d
10							25.6 d	32 d	38.4 d
11							23.3 d	32 d	37.8 d
12							21.3 d	32 d	32 d
13							19.7 d	32 d	34.5 d
14							18.3 d	32 d	32 d
15							17.1 d	32 d	32 d
16							16 d	16 d	16 d

Besides that, we can also learn from Table 7.4 and the corresponding charts that in both cases, CCI and CCI-LT, some frequencies are dominated by smaller ones, i.e. the same level of system inventory can be reached with fewer transports. In the table the dominated values are underlined. In case of the heijunka levelled system, there even exist cases in which a higher frequency leads to a higher total inventory. In Table 7.4, they are underlined twice. Hence, in contrast to the CCI case the inventory cost functions are not convex over the integer domain $f \in [1, .., T]$. Please note that a formal proof of the convexity in the CCI case is given in (Kovalev and Ng 2008).

Remember from Chapter 3 that Chuah and Yingling (2005) reported that for a planning horizon of one day with a maximal frequency of eight, only the frequency set $\{1, 2, 3, 4, 8\}$ is allowed at a Toyota Motor Manufacturing plant in the US. That means, Toyota ignores all dominated frequencies. However, due to better consolidation possibilities resulting from smaller shipping volumes a dominated frequency might be of interest to reduce total operational cost.

In case of the CDI model, the inventory is not equally distributed between supplier and plant if f is not a divisor of T. This needs to be taken into account if restricted storage space determines minimum frequencies. An example is given in Table 7.5 assuming that the applied patterns also minimize the inventory at the supplier. In case of a pull based policy, the inventory level at the supplier would correspond to the level in the plant and vice versa.

Table 7.5: Inventory in the CDI case depending on the demand rate d for a system cylce time $T = 8$ and different frequencies f assuming a transport time of $tt = 0$

frequency	$\mathcal{I}_{supp}$	$\mathcal{I}_{plant}$	$\mathcal{I}_{total}$
4	$8\,d$	$8\,d$	$16\,d$
5	$7\,d$	$9\,d$	$16\,d$
6	$6\,d$	$10\,d$	$16\,d$
7	$5\,d$	$11\,d$	$16\,d$

7.7.2 Volume Consistency

Depending on the length of the interval without shipment, the volumes vary in the CDI model without levelling mechanism. This might lead to an unlevelled use of transport resources. Therefore, in lean systems often the frequencies and the planning horizon are restricted to be a multiple of two times the base period (see (Monden 2012)) to reach the constant cycle property for time discrete models. However, such a strategy is difficult to apply if a natural planning horizon is not a power of two, for example a working week of five days. Further, such a strategy restricts the consolidation possibilities during routing.

In contrast, constant-cylce systems and heijunka levelled CDI-LT systems exhibit levelled transport volumes and ensure a levelled use of the transport capacity in a deterministic environment. However, there exists a difference: In case of varying demands, the heijunka mechanism caps the transport volume, while in the constant cycle system the volumes might strongly vary.

7.7.3 Schedule Compatibility and Differing Opening Hours

Besides the fact that delivery profile schedules resulting from applying the CDI or CDI-LT system to weekly recurring milk runs are easy to memorise, the schedules have another big advantage: They are compatible with opening hours of the logistics departments of the suppliers. If arrival times are equally distributed over the course of the week in a constant cycle system, there are usually conflicts with opening hours and night drive bans. The same is true for multi echelon transport networks: The schedules must be consistent with the opening hours of transshipment points or consolidation centres.

In daily recurring milk runs with kanban cycles, opening hours influence the lot sizes and the inventory behaviour significantly: In the example of Figure 7.15, it is assumed that the supplier runs one shift per day and the plant three shifts, and that two equally distributed pick ups per shift at the supplier site are requested. In an unlevelled kanban case, the volumes strongly deviate for the following reason: The time between the second pick up (of the day before) and the first pick up contains the two shifts in which the plant consumes parts while the supplier is closed plus half of the shift in which the supplier is open. However, the inter arrival time between the first and the second pick up (of the same day) is just half a shift. In this case, heijunka levelling might be preferable since, besides less deviating shipping volumes, it also leads to lower inventory levels.

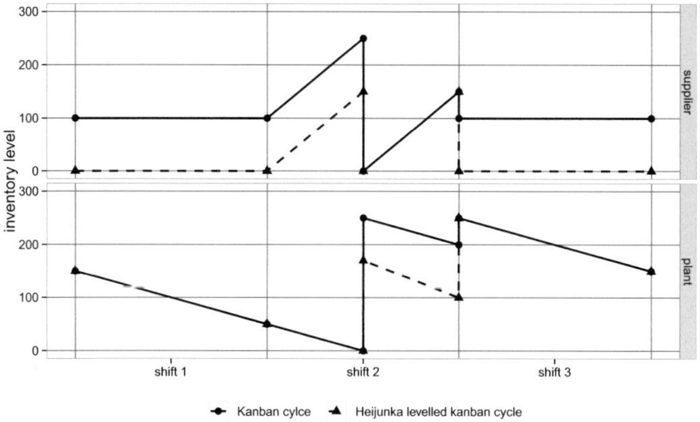

Figure 7.15: Inventory level at the supplier (first row) and at the plant (second row) for an unlevelled kanban cycle and a heijunka levelled one if the opening hours of the supplier are only during the second shift of the day, while the plant is open 24 hours

7.8 Summary

In this chapter we shortly introduced existing finite horizon inventory models with continuous time and models assuming that orders are fulfilled at discrete steps in time. These models are proposed in literature for push based order systems.

Based on these finite horizon models, we introduced unified cyclic versions: The cyclic continuous time inventory model (CCI) and the cyclic discrete time inventory model (CDI). The continuous time models result in constant cycle schedules, while the discrete time counterparts result in transport patterns or delivery profiles.

Both models are a generalisation of their finite equivalents since the first shipment does not need to be scheduled at the beginning of the first period. We proposed formulas to determine the inventory at the plant, in transit and at the supplier. Further, we derived optimality properties for patterns which yield minimum total inventory levels over the course of the planning horizon for a given frequency.

For the CDI we showed that the inventories are not symmetrically distributed between the supplier and the plant: In a pull based system, more inventory is stored at the point of consumption. This is beneficial from a supply chain management point of view. In a push based system, it is the other way around. For CCI the inventories are distributed symmetrically, and no differences occur for pull and push based policies.

Table 7.6: Characteristics of different introduced kanban systems

CCI	CDI	CDI-LT
$\mathcal{I}_{tot} = \frac{T}{f} \cdot T \cdot d$	$\mathcal{I}_{tot} = \lceil \frac{T}{f} \rceil \cdot T \cdot d$	$\mathcal{I}_{tot} = \frac{T - gcd(T, f) + f}{f} \cdot T \cdot d$
minimum resulting inventories	–	–
levelled transport lot sizes	–	levelled (capped) transport lot sizes
–	easy to memorise schedules	
–	compatible with all common shift types of suppliers	
–	arrival times compatible in multi echelon transport networks	
–	–	minimal inventory in case of strongly deviating opening hours

Since varying transport lot sizes in case of delivery profile schedules clash with the goal of a levelled use of transport resources, we proposed a third model: A cyclic discrete time model with levelled transport lots (CDI-LT). We verbally described an optimality property for patterns yielding minimum total inventory and proposed an efficient algorithm to generate all patterns fulfilling this property. We furthermore proposed formulas to determine the inventory at the plant and the supplier. In terms of inventory symmetry, the CDI-LT shows the same properties as CCI: The inventory is symmetrically distributed between the supplier and the plant, and there exist no differences between pull and push based policies.

To complete this chapter, we discussed the characteristics of the respective cyclic order strategies. They are summarized in Table 7.6.

8 Milk Run Scheduling

Regularity and consistency are important features of milk run schedules. Therefore, the goal of this chapter is to introduce different possible configurations of milk run scheduling (MRS) models considering these aspects. For this chapter we assume that the suppliers assigned to the milk run concept are known. The task of these models is to assign a weekly pattern to every supplier and to determine and to schedule the milk run tours. In Chapter 9, this formulation is extended to integrate the transport concept assignment.

The milk run scheduling models consider the inventory order strategies corresponding to the models CCI, CDI, and CDI-LT proposed in the preceding chapter. The strategies can be used for both, push and pull based policies. For an easier understanding, we focus in this chapter on the kanban case. However, to stay general, we point out whenever there appear differences between push and pull policies.

Since our cyclic inventory models assume that the optimal – constant or varying – inter arrival times can be guaranteed in a milk run schedule for every single supplier, we need to consider them during the routing part. This is reached in two steps: In our PVRP-SC model, only inventory optimal (weekly) patterns for our order strategies are allowed using the optimality properties derived in the preceding chapter. This only assures that the correct days are chosen for a pickup, but it does not restrict the time during the day. Therefore, we extend the model imposing additional scheduling constraints to assure that the deviation from the optimal inter arrival time is capped. This is important since a deviation of, e.g., half a day would influence the inventory levels, while a deviation of 1 to 2 hours, usually, does not influence the inventory behaviour significantly.

Hence, our model has macro periods – typical for lot sizing models – and micro periods, in which the typical VRP modelling in terms of time windows can be applied. This formulation allows to easily consider forbidden periods or pick up day combinations. In case of the CCI order policy, we impose scheduling constraints assuring inter arrival time consistency (ITC), and in the other two cases (CDI and CDI-LT) the constraints ensure arrival time consistency (ATC).

Please note that we assume, according to our goals formulated in Chapter 6, that the base period is one day and the planning horizon is one or two weeks. The models are also applicable to cases with a different period length, for example half a day or a shift. However, in these cases the concept of arrival time consistency does not necessarily make sense.

In the first section of this chapter, we introduce the models considering the different order strategies and scheduling requirements, as well as driver regularity (see also Section 2.4). In order to be able to evaluate these models on small instances, we introduce an a priori column generation approach. The results in terms of solution quality and run times are presented before we conclude this chapter with a short summary.

8.1 Milk Run Scheduling Models

As already mentioned in the introduction of this chapter, we propose a PVRP-SC along with the cyclic inventory models for the milk run scheduling decision: This model variant can be considered as a decomposed inventory routing problem. Since there is a strong focus on regularity and reliability in both flow of parts and scheduling, this decomposition combined with scheduling requirements seems to be adequate. The SDVRP models, which are also proposed in literature for milk run scheduling, do not allow for an integrated milk run tour determination and scheduling and, hence, are not appropriate for the weekly use case.

Vehicle routing problems are often presented as arc flow formulations, which can be directly implemented and solved using a MIP solver. However, an arc based PVRP-SC (see for example (Rusdiansyah and Tsao 2005) or (Francis et al. 2008)) leads to prohibitively long run times, even for small instances. Therefore, we directly introduce a path based formulation. In this kind of VRP formulations, it is assumed that all feasible tours can be enumerated a priori or iteratively within a so called column generation framework.

The latter is a successful exact method for solving VRP, which alternates between the solution of the relaxed path flow model (master problem) and a column generation sub procedure using information of the master problem (see for example (Mourgaya and Vanderbeck 2007)) in order to avoid to enumerate all possible tours. However, the approaches are usually tailored to specific problem classes. Thus, in this thesis we propose a simple a priori column generation procedure, similar to the one proposed in (Peterson et al. 2010), in order to be able to evaluate many different model configurations on small and mid sized instances.

8.1.1 Base Model

Table 8.1 describes in detail all sets, parameters and decision variables of our milk run scheduling model. The term "pattern" – or delivery profile – is used for the allowed pick up day combinations in order to avoid confusions with the term "schedule", which refers in the following to the whole solution.

Let us assume so far that the set $\mathcal{R}$ contains all routes, feasible with respect to the maximum allowed tour duration D^T and the maximum vehicle capacity Q. The route generation procedure is discussed in Section 8.2.

All milk run schedule variants contain the MRS core model. It corresponds to the path flow model of a classic PVRP and is extended in the following sections by different aspects which are characteristic for the milk run scheduling problem.

$$\text{(MRS)} \qquad \min \sum_{t \in \mathcal{R}} \sum_{d \in \mathcal{T}} \sum_{k \in \mathcal{K}} c_t \cdot z_{tdk} + \sum_{i \in \mathcal{V}^c} \sum_{r \in \mathcal{C}_i} c^h_{ir} \cdot u_{ir} \tag{8.1}$$

subject to

$$\sum_{r \in \mathcal{C}_i} u_{ir} = 1 \quad i \in \mathcal{V}^c \tag{8.2}$$

$$\sum_{t \in \mathcal{R}} z_{tdk} \leq 1 \quad d \in \mathcal{T}, k \in \mathcal{K} \tag{8.3}$$

$$\sum_{r \in \mathcal{C}_i} u_{ir} \cdot \alpha_{rd} - \sum_{t \in \mathcal{R}} \sum_{k \in \mathcal{K}} z_{tdk} \cdot a_{ti} = 0 \quad i \in \mathcal{V}^c, d \in \mathcal{T} \tag{8.4}$$

$$z_{tdk} \in \{0,1\} \quad t \in \mathcal{R}, d \in \mathcal{T}, k \in \mathcal{K} \tag{8.5}$$

$$u_{ir} \in \{0,1\} \quad i \in \mathcal{V}^c, r \in \mathcal{C}_i \tag{8.6}$$

With constraint set (8.2) we ensure that every supplier i is assigned to one of its feasible patterns r. Constraint set (8.3) assures that every vehicle k is assigned to at most one tour t in period d while line (8.4) links the pattern and the tour assignment decision. Whenever a pattern r is chosen for supplier i containing the period d ($\alpha_{rd} = 1$), there must be exactly one tour t containing supplier i ($a_{ti} = 1$) assigned to one vehicle available at period d. The last two lines restrict the variables z_{tdk} and u_{ir} to be binary.

The decisions are taken in a way that the total transport cost for the selected tours and the inventory holding cost for the assigned pattern are minimized (see (8.1)).

The auxiliary variable z_{td} indicating if tour t is driven in period d is used in different extensions and is defined as follows:

$$\sum_{k \in \mathcal{K}} z_{tdk} = z_{td} \quad t \in \mathcal{R}, d \in \mathcal{T} \tag{8.7}$$

$$z_{td} \in \{0,1\} \quad t \in \mathcal{R}, d \in \mathcal{T} \tag{8.8}$$

Please note that this model can be easily extended to the case with a heterogeneous fleet containing vehicles with different capacity and cost parameters (see Chapter 10).

8.1.2 Modelling Kanban Systems and Service Choice

In case of a heijunka levelled system (HEI), in which the frequency is fixed, the pick up volumes are known in advance. Consequently, the capacity constraints can be completely checked during the a priori route generation. In this case, only capacity feasible solutions are added to set $\mathcal{R}$.

However, in case of an unlevelled lot size system and, if the frequency is not fixed in advance, the pick up volumes depend on the pattern decisions. Hence, in the a priori route generation we can only check capacity

Table 8.1: Overview of sets, parameters, and variables

Sets and parameters	
$\mathcal{R}$	set of all feasible routes considering the maximum tour length D^T and the vehicle capacity Q with index t
$\mathcal{V}$	set of n supplier nodes including 0 representing the start node of the depot and $n+1$ representing the end node in the depot with index i, j
$\mathcal{V}^c$	set of supplier nodes with $\mathcal{V}^c = \mathcal{C} \setminus \{0, n+1\}$
$\mathcal{C}_i$	set of feasible patterns for customer i with index r
$\mathcal{K}$	set of available vehicles with index k
$\mathcal{T}$	set of periods (days) in the planning horizon (system cycle time) with index d
D^P	duration of the periods
α_{rd}	equals 1 if and only if day d belongs to pattern r else 0
$\mathcal{T}_r$	ordered set of periods, containing all days of service of a pattern r with $\mathcal{T}_r = \{d \in \mathcal{T} \mid \alpha_{rd} = 1\}$
a_{ti}	equals 1 if and only if supplier i belongs to route t else 0
c_t	transport cost for route t
c^h_{ir}	inventory holding cost for supplier i for pattern r
sos_{ti}	earliest start of service of supplier i in route t assuming a starting time of the route at the beginning of a period at time $t = 0$
d_{ird}	demand at supplier i on day d for schedule r, 0 on days without pick up, d^{KAN}_{ird} is used for the unlevelled kanban case, d^{HEI}_{ird} for the heijunka levelled case given in a common unit, e.g. pallets or weight
Q	vehicle capacity
D^T	maximal allowed tour duration
Variables	
z_{tdk}	equals 1 if vehicle k services route t on day d else 0
z_{td}	equals 1 if route t is serviced on day d else 0
u_{ir}	equals 1 if schedule $r \in \mathcal{C}_i$ is assigned to customer i else 0
w_{ik}	equals 1 if supplier i is serviced by vehicle k else 0
w^{pen}_{ik}	equals 1 if supplier i is serviced by another vehicle k than before else 0
s^{ser}_{id}	start of service at supplier i on day d if at day d supplier i is visited else 0
s^{tour}_{td}	start of tour t on day d if tour t is serviced on day d else 0
v_{ikd}	equals the pick up volume of supplier i if at day d a pick up by vehicle k is scheduled else 0

constraints with regard to a minimum volume for every customer. In these cases, the capacity constraint must be additionally handled within the path flow model.

Capacity Restrictions in a Kanban System

For kanban system (KAN) which is not levelled the parameter d_{ird}^{KAN} is introduced. It represents the pick up volume for a supplier on day d if pattern r is chosen. The volume picked up at a supplier i on day d by vehicle k is represented by the additional variables v_{ikd}. Hence, we can model capacity restrictions as follows:

$$\sum_{r\in\mathcal{C}_i} d_{ird}^{KAN} \cdot u_{ir} - Q \cdot (1 - \sum_{t\in\mathcal{R}} z_{tdk} \cdot a_{ti}) \leq v_{ikd} \quad i \in \mathcal{V}^c \tag{8.9}$$

$$\sum_{r\in\mathcal{C}_i} d_{ird}^{KAN} \cdot u_{ir} + Q \cdot (1 - \sum_{t\in\mathcal{R}} z_{tdk} \cdot a_{ti}) \geq v_{ikd} \quad i \in \mathcal{V}^c \tag{8.10}$$

$$\sum_{i\in\mathcal{V}^c} v_{ikd} \leq Q \quad k \in \mathcal{K}, d \in \mathcal{T} \tag{8.11}$$

$$v_{ikd} \geq 0 \quad i \in \mathcal{V}^c, k \in \mathcal{K}, d \in \mathcal{T} \tag{8.12}$$

Constraint sets (8.9) and (8.10) link the new volume variables to the schedule decision, whereas set (8.11) restricts the volume. The volume for a visit is greater than 0 for vehicle k in period d if the visit is part of the tour assigned to vehicle k.

If the service frequency is given in advance, all patterns in $\mathcal{C}_i$ correspond to the preselected frequency. If service choice is allowed, all patterns meeting a minimum frequency form the set of feasible patterns. So, all service choice cases can be represented by the same models: a fixed service choice (FIXSC), a service choice fulfilling at least a minimum frequency (ALSC), and a free service choice (FRSC), which allows all frequencies for all suppliers.

Capacity Restrictions in a Heijunka Levelled Kanban System

If service choice (ALSC or FRSC) is considered, also in the heijunka levelled case and in case of constant inter arrival times, the volumes depend on the frequency of the chosen pattern. In line with the formulation above, a parameter d_{ird}^{HEI} is introduced, representing the pick up volume for a supplier on a day d if schedule r is chosen. Then, d_{ird}^{KAN} in constraints (8.9) and (8.10) can simply be replaced by this new parameter.

8.1.3 Modelling Scheduling Requirements

Coming to the scheduling aspect of the problem, two cases can be differentiated from a modelling point of view: The start time for all tours is the same for all days (FIX), or the tour start time is free (FR). In the first case, it is not necessary to introduce any new variables. However, since (FIX) is a special case of (FR), we only introduce a slightly more complex model based on new variables.

We add two sets of continuous, positive variables, one representing the start of a tour t on a certain day d (s_{td}^{tour}), and one representing the start of service at a supplier i on day d (s_{id}^{ser}). The start of service of supplier i within route t assuming a tour start at the beginning of the day (in our case 0) is given by parameter sos_{ti} and is a result of the route generation phase. Since we do not allow waiting time, the start of service of a supplier i can simply be expressed as the sum of the start of the tour in which i is serviced plus the earliest start of service for i on this tour.

$$sos_{ti} + s_{td}^{tour} + D^P \cdot (1 - z_{td} \cdot a_{ti}) \geq s_{id}^{ser} \quad i \in \mathcal{V}^c, \; t \in \{\mathcal{R} | a_{ti} = 1\}, \; d \in \mathcal{T} \tag{8.13}$$

$$sos_{ti} + s_{td}^{tour} - D^P \cdot (1 - z_{td} \cdot a_{ti}) \leq s_{id}^{ser} \quad i \in \mathcal{V}^c, \; t \in \{\mathcal{R} | a_{ti} = 1\}, \; d \in \mathcal{T} \tag{8.14}$$

$$\left(D^P - sos_{t(n+1)}\right) \cdot z_{td} \geq s_{td}^{tour} \quad t \in \mathcal{R}, d \in \mathcal{T} \tag{8.15}$$

$$0 \leq s^{ser}_{id} \leq D^P \quad i \in \mathcal{V}^c, d \in \mathcal{T} \tag{8.16}$$

$$0 \leq s^{tour}_{td} \leq D^P \quad t \in \mathcal{R}, d \in \mathcal{T} \tag{8.17}$$

The constraint sets of lines (8.13) and (8.14) assure that the start of service of supplier i corresponds to the described sum, but only if supplier i is serviced on tour t at period d. Otherwise, s^{ser}_{id} is pushed to 0. In the next line, it is assured that the tour ends within the period in which it is started by allowing only start times which are smaller than the length of the period minus the return time to the depot (represented by parameter $sos_{t(n+1)}$). This restriction is reasonable if the period is, for example, a day and the maximum allowed driving time is considerably smaller – as assumed in our small examples of the next section. If the period is, for example, half a shift and if the driving times might be longer, this restriction should be dropped. In (8.16) and (8.17) both new variables are restricted to be between the start and the end time of a period corresponding in our case to 0 and the length of the period D^P, respectively.

Since in many cases a lot of different tour start times and, furthermore, a lot of combinations of different tour start times lead to the same (optimal) solution, we break symmetry by favouring early start times. Therefore, the objective function (8.1) is modified by adding a third term:

$$\gamma \sum_{t \in \mathcal{R}} \sum_{d \in \mathcal{T}} s^{tour}_{td}$$

For the parameter γ a sufficiently small value is chosen so that it does not influence the total cost.

If time window constraints for suppliers exist, they can be imposed on the start of service variables. However, if at the same time the start of the tour is fixed, it is more efficient to consider time window constraints during the a priori route generation process since they restrict the search space and limit the number of feasible tours.

Arrival Time Consistency

In an arrival time consistent schedule, the arrival times for all visits at the same supplier may differ by a maximum of L time units. Using the new scheduling variables, this can be reached by imposing the following constraints:

$$-L - D^P \cdot (1 - u_{ir}) \leq s^{ser}_{id_1} - s^{ser}_{id_2} \quad i \in \{\mathcal{V}^c | f_i > 1\}, r \in \mathcal{C}_i, \\ d_1, d_2 \in \{\mathcal{T} | \alpha_{rd_1} = 1 \\ \wedge \alpha_{rd_2} = 1 \wedge d_1 < d_2\} \tag{8.18}$$

$$L + D^P \cdot (1 - u_{ir}) \geq s^{ser}_{id_1} - s^{ser}_{id_2} \quad i \in \{\mathcal{V}^c | f_i > 1\}, r \in \mathcal{C}_i, \\ d_1, d_2 \in \{\mathcal{T} | \alpha_{rd_1} = 1 \\ \wedge \alpha_{rd_2} = 1 \wedge d_1 < d_2\} \tag{8.19}$$

Since the absolute difference between two arrival times needs to be smaller than L, we introduce two constraint sets to linearise the absolute value function. For every feasible schedule of every supplier the constraints are defined for all pairs of periods which are part of the schedule. The constraints only take effect if a schedule r is chosen for a supplier i. This formulation is similar to the ones proposed in (Groër et al. 2009) and (Feillet et al. 2010), with the difference that the latter ones only consider fixed periods of service – corresponding to a fixed u_{ir} in our model.

If all tours should start at the same time of the day (FIXATC), we set all s^{tour}_{td} variables to the corresponding start time. If the start time is free (FRATC), these variables are not further restricted.

Inter Arrival Time Consistency

To harmonize the inter arrival times (FRITC) between two consecutive visits to the same supplier, we introduce an ordered set $\mathcal{T}_r$, which contains all days of service of the schedule r in an increasing order. Furthermore, we define the function $at(set, n)$, which returns the element of an ordered set on position n. $|\mathcal{T}|$ stands for the number of periods of the system cycle time and $f_r = \sum_{d \in \mathcal{T}} \alpha_{rd}$ for the frequency represented by pattern r. Therewith, we can impose the following constraints:

$$s^{ser}_{id_2} - s^{ser}_{id_1} + (d_2 - d_1)D^P - M(1 - u_{ir}) \leq \frac{|\mathcal{T}| \cdot D^P}{f_r} + L$$
$$i \in \{\mathcal{V}^c | f_i > 1\}, r \in \mathcal{C}_i, d_1 = at(\mathcal{T}_r, n),$$
$$d_2 = at(\mathcal{T}_r, n+1), n \in \{1..f_r\} \quad (8.20)$$

$$s^{ser}_{id_2} - s^{ser}_{id_1} + (d_2 - d_1)D^P + M(1 - u_{ir}) \geq \frac{|\mathcal{T}| \cdot D^P}{f_r} - L$$
$$i \in \{\mathcal{V}^c | f_i > 1\}, r \in \mathcal{C}_i, d_1 = at(\mathcal{T}_r, n),$$
$$d_2 = at(\mathcal{T}_r, n+1), n \in \{1..f_r\} \quad (8.21)$$

$$s^{ser}_{id_1} - s^{ser}_{id_2} + (|\mathcal{T}| + d_1 - d_2)D^P - M(1 - u_{ir}) \leq \frac{|\mathcal{T}| \cdot D^P}{f_r} + L$$
$$r \in \mathcal{C}_i, i \in \{\mathcal{V}^c | f_i > 2\}, d_1 = at(\mathcal{T}_r, 0), d_2 = at(\mathcal{T}_r, f_i) \quad (8.22)$$

$$s^{ser}_{id_1} - s^{ser}_{id_2} + (|\mathcal{T}| + d_1 - d_2)D^P + M(1 - u_{ir}) \geq \frac{|\mathcal{T}| \cdot D^P}{f_r} - L$$
$$r \in \mathcal{C}_i, i \in \{\mathcal{V}^c | f_i > 2\}, d_1 = at(\mathcal{T}_r, 0), d_2 = at(\mathcal{T}_r, f_i) \quad (8.23)$$

The first two constraint sets (8.20) and (8.21) control the inter arrival times between all pairwise consecutive visits of every feasible schedule for every supplier with a frequency request greater than 1. The last two constraints (8.22) and (8.23) control the inter arrival time between the last and the first visit of every feasible schedule. If the frequency is just 2, the inter arrival times are harmonized across system cycles automatically by just restricting one of the two inter arrival times.

Since s^{ser}_{id} represents a point in time within the period d, we add $(d - 1) \cdot D^P$ to these variables in order to get the actual time difference across periods (or days). In case of the second pair of constraints, we also add $|\mathcal{T}| \cdot D^P$ to the first visit since this is the first visit of the second system cycle. As in the section before, the constraints only take effect if the schedule r is assigned to supplier i. M stands for a sufficiently big number, which can be replaced by $D^P \cdot |\mathcal{T}|$.

On the right hand side, the fraction represents the optimal inter arrival time for the frequency of the chosen schedule, from which the absolute time difference is allowed to differ only by L. It does not make sense to limit the tour start times in this case since the arrival times are distributed across the whole day: A fixed start time – for example at the beginning of the day – would make the scheduling problem in many cases infeasible, as a supplier cannot be reached at the end of the day without violating the maximum tour duration.

In both time consistency cases, in which we jointly solve the routing and the scheduling part, the constraints we add to the base model are linking constraints representing the interdependence between tours through dynamic time windows (see (Drexl 2012)). For VRP considering similar types of dynamic time windows, Andersson et al. (2011) also proposed an a priori column generation approach.

8.1.4 Modelling Driver Regularity

Regarding driver regularity, we consider three different configurations assuming that vehicle and driver are a fixed pair: In a driver consistent (DC) schedule, all visits at one supplier are serviced by the same vehicle (see Section 2.4), whereas in a partial driver consistent (PDC) schedule the number of driver changes is minimized. A model without any requirement on driver regularity (NODC) is considered as well.

Driver Consistency

For modelling total driver consistency we introduce additional binary variables w_{ik}, indicating by 1 the vehicle k which services supplier i (0 otherwise), and impose the following constraint sets (compare also the formulations of Groër et al. (2009) and Coelho et al. (2012)):

$$\sum_{k \in \mathcal{K}} w_{ik} = 1 \quad i \in \mathcal{V}^c \tag{8.24}$$

$$\sum_{t \in \mathcal{R}} z_{tdk} \cdot a_{ti} \leq w_{ik} \quad i \in \mathcal{V}^c, d \in \mathcal{T}, k \in \mathcal{K} \tag{8.25}$$

$$w_{ik} \in \{0,1\} \quad i \in \mathcal{V}, k \in \mathcal{K} \tag{8.26}$$

In line (8.25) exactly one vehicle k is assigned to every supplier i, while with constraint set (8.24) it is assured that supplier i can only be visited by the assigned vehicle k.

Partial Driver Consistency

If driver consistency is not necessary due to organisational reasons, a partial driver consistency might be beneficial. In this formulation, having different drivers (vehicles) who visit the same supplier is penalised in the objective function. Therefore, a second set of binary variables w_{ik}^{pen} is introduced to indicate if a supplier is visited by a vehicle different from the one which performed the first visit. This concept was introduced in (Coelho et al. 2012), and the following constraints are very similar to the original formulation:

$$\sum_{k \in \mathcal{K}} w_{ik} = 1 \quad i \in \mathcal{V}^c \tag{8.27}$$

$$\sum_{t \in \mathcal{R}} z_{tdk} \cdot a_{ti} \leq w_{ik} + w_{ik}^{pen} \quad i \in \mathcal{V}^c, d \in \mathcal{T}, k \in \mathcal{K} \tag{8.28}$$

$$w_{ik}, w_{ik}^{pen} \in \{0,1\} \quad i \in \mathcal{V}, k \in \mathcal{K} \tag{8.29}$$

In line (8.27) we assign supplier i to a first visiting vehicle, whereas, with the next set of constraints, we push the variable w_{ik}^{pen} to one for every extra vehicle servicing the same supplier i. The objective function (8.1) is then extended by a penalising term:

$$\beta \sum_{i \in \mathcal{V}^c} \sum_{k \in \mathcal{K}} w_{ik}^{pen}$$

The parameter β allows to control the importance of driver consistency compared to the operational cost.

8.2 A Priori Route Generation

We chose a simple approach for generating all feasible routes for a set of given customers which is computationally not prohibitive in case of the instance sizes we use for our experiments. Actually, solving the path flow model with its extensions is the time consuming process.

As already mentioned, the maximal tour duration (D^T) as well as the capacity (Q) restrictions – based on the lower bound of shipping volumes – need to be considered. For the different order strategies the calculation of the lower bound of the transport lots differs: In the unlevelled case, the smallest inter arrival time over all feasible patterns defines the volume, in the levelled case, the smallest feasible frequency.

For enumerating the feasible tours, first of all, we calculate the maximal number of supplier stops M^{stop} a route can contain due to capacity restrictions. To this end, we sort the visits by their lower volume bounds in an increasing order and check how many visits fit at most in a vehicle without exceeding its capacity. Then we generate all subsets of $\mathcal{V}^c$ of size 1 through M^{stops} not considering the depot and check for each subset if the capacity constraint is fulfilled.

As mentioned in Section 8.1.3, time consistency causes a dependency between tours: If one route is selected, the arrival times for all following visits of suppliers which are on this route are restricted. These dynamic time windows might induce that the shortest tour with regard to distance and driving time for a set of suppliers on succeeding days is no longer feasible. A constructed example of such a case is shown in Figure 8.1: If the extra time to include customer, who does not need to be visited every day, into a route is longer than L, a route not optimal with regard to distance and time for this cluster of suppliers is part of the optimal solution.

For the a priori route generation this means that not only the shortest routes with regard to their tour length are added to the set of feasible routes, but all routes of every generated set which do not exceed the maximal allowed tour duration D^T. The number of feasible routes or columns grows quickly with the number of visits[1] and, therefore, this

[1] An upper bound can be given by: $\sum_{k=1}^{M^{stops}} \binom{n}{k} \cdot k!$

approach is only tractable for small instances. However, since the route generation procedure for our small examples only takes between a few seconds and 2 minutes, we omit to develop a more sophisticated approach – for example a tree constructing algorithm with dominance rules as proposed in (Andersson et al. 2011).

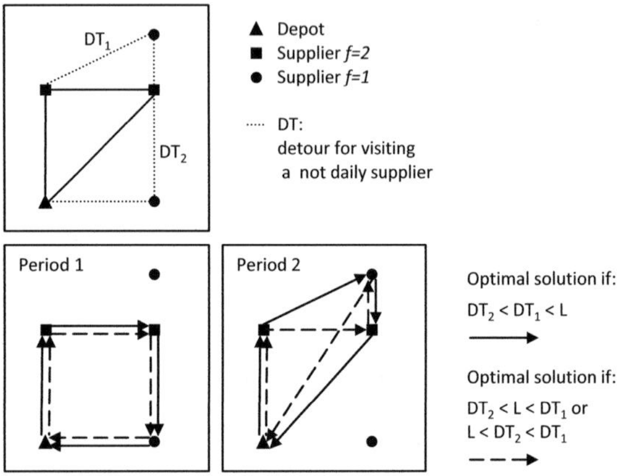

Figure 8.1: Solutions for a small example of two periods, one vehicle and four suppliers to show the effect of dynamic time windows

Beside the set of feasible routes, the cost and the earliest start of service for all visits – including the arrival time at the depot – are input parameters for the set cover model.

8.3 Computational Experiments

In this section we introduce experiments that are designed to show the impact of different formulations on the resulting plans. We solve a set of 10 small instances with 8 suppliers, serviced on 5 days, resulting in up to 32 visits per schedule. A summary of all results and a short discussion follows in the last subsection.

8.3.1 Instances and Model Configurations

We generated 10 random test instances using the following assumptions and parameters closely resembling the small ConVRP problems of Groër et al. (2009):

- Node locations are uniformly distributed on a square with vertices $(0,0)$, $(10,0)$, $(10,10)$, $(0,10)$ with the depot in vertex $(0,0)$. The travel distances and travel times are calculated by means of the Euclidean norm and are assumed to correspond to the cost.
- The planning horizon has 5 periods, all frequencies between 1 and 5 are feasible. The respective probabilities to select a minimum frequency for a supplier are as follows: $(1 : 0.25), (2 : 0.2), (3 : 0.1), (4 : 0.2), (5 : 0.25)$
- Only patterns resulting in minimal inventory for the given frequency and for an order strategy are considered as feasible (see Chapter 7).
- Demand rates are uniformly distributed on $\{1, 2, 3\}$, and we assume that all goods are divisible, i.e. for the sake of simplicity, we ignore any rounding issues occurring in the heijunka case.
- The maximum travel time and the length of a shift is $D = 30$, and the vehicle capacity is $Q = 10$.
- The length of a day is $D^P = 90$, which is three times the shift length.
- The standard maximum arrival time difference is $L = 5$. For sensitivity studies we also use $L = 0$ and $L = 2$.
- The standard number of available vehicles and drivers per day is 2. For some cases we only received feasible solutions allowing more vehicles per day.

Please note that we do not explicitly consider inventory holding cost even if it is possible in the proposed formulation. However, in practice, it is difficult to quantify the positive effects of high frequencies, especially, as there is an interdependence between frequency decisions and for example the reserved area and number of specialized containers for a certain item.

It seems more natural to give a minimum frequency instead of estimating the monetary advantage value of a better visibility or a smoother flow. However, the milk run scheduling model can be used for supporting this frequency assignment decision by applying the FRSC configuration and calculate a lower bound for transport cost.

For an easier overview, we classify the different model variations by the categories order strategy, service choice type, scheduling type, and driver regularity (see Table 8.2). Since the number of possible variants is high, we conducted the experiments in four phases excluding step by step combinations which – in our judgement – do not contribute to gain new insights in the next phase. Each phase is dedicated to show the impact of the model on the resulting schedule with regard to a special issue.

Table 8.2: Overview of model variants per category

Order strategy	Service choice	Scheduling	Driver regularity
KAN	FIXSC	FIXATC	DC
HEI	ALSC	FRATC	PDC
	FRSC	FRITC	NODC
		NOTC	

Furthermore, we describe for every phase the impact on the run times of the path flow model. From this behaviour one can draw conclusions of the performance of an – exact or heuristic – column generation approach and of set cover based heuristics (as proposed for example in (Peterson et al. 2010)).

8.3.2 Phase I: Kanban vs. Heijunka

The main focus in this phase is to check if the objective values differ for pure kanban controlled and heijunka levelled milk run systems in case of varying inter arrival times. More precisely, we investigate if the different splits of shipments lead to different chances of finding compatible loads for neighbouring suppliers.

Therefore, we compared the results of three configurations of a pure and levelled kanban system for our 10 instances. The first configuration, without any consistency requirement, serves as a lower bound for the driving distance and the number of vehicles and is used as a baseline throughout all phases. We furthermore compare arrival time consistent schedules with free starting times with and without driver consistency (see also Table 8.3). Remember that inter arrival time consistent schedules automatically lead to levelled volumes and are not considered in this phase, but in phase III studying the effect of different scheduling types.

Table 8.3: Model configurations of phase I

KAN	HEI
FIXSC NOTC NODC	FIXSC NOTC NODC
FIXSC FRATC DC	FIXSC FRATC DC
FIXSC FRATC NODC	FIXSC FRATC NODC

Impact on Solution Quality

The impact on the driving distance and the number of vehicles is small: Only in 2 out of 10 instances – but then in all three configurations – there are differing total driving distances between the pure kanban and the levelled case. However, the absolute differences are rather small (between 0.3 to 3.0%), and they deviate in both directions: For one instance all three heijunka configurations have slightly shorter driving distances, for the other instance it is the other way around. The differences between configurations with driver consistency are a little bit smaller (see Table 8.4).

For just one instance in just one configuration (FRATC NODC), the number of tours differs: In the heijunka case with its equal split of volumes, a tour is infeasible, which in the kanban case is feasible since for two of the corresponding suppliers the low pick up volumes are assigned to this tour. In the heijunka case then one extra tour becomes necessary.

Table 8.4: Overview of deviation of the objective value of KAN compared to HEI for fixed service choice (FIXSC) configurations aggregated over all instances

Model configuration	Deviation of objective value KAN vs. HEI			Instances without deviation
	mean	min	max	
NOTC NODC	0.01%	-2.9%	3.0%	80%
FRATC DC	0.19%	-0.3%	2.2%	80%
FRATC NODC	0.00%	-3.0%	3.0%	80%

However, in general we can state that it has no big impact on the quality of the resulting schedules if heijunka levelling is applied or not. Even for our small instances, for which the chance to find compatible loads is naturally smaller, the differences are small.

Impact on Run Times

Even if there are no big differences in terms of the quality of the resulting schedules, from an algorithmic point of view the unlevelled variant tends to be more difficult to solve. As already described, the shipping volumes depend on the concrete schedule and, therefore, the volumes are modelled as a decision variable, and capacity constraints need to be introduced in the path flow model. As a result, the average and median run times increase (see Figure 8.2).

Compared to the KAN configuration, the run times in HEI increase on average by a factor of 4.1 with a median of 2.5 in the NOTC NODC case. For the KAN FRATC DC configuration the factor even goes up to 18, for the KAN FRATC NODC to 13, whereas the median in both cases stays below 1.5. That shows that there exist a couple of extreme outliers with a run time up to 100 times longer than in the heijunka case. That means, run times for the kanban configuration are rather unstable.

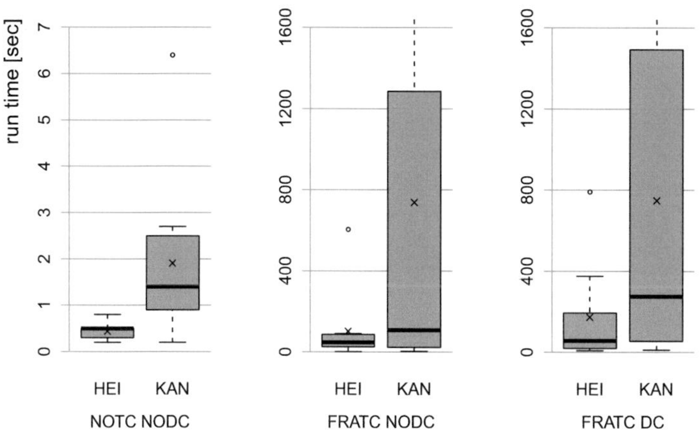

Figure 8.2: Boxplots with the mean value represented as a cross comparing run times for solving different KAN and HEI configurations assuming a fixed service choice (FIXSC)

8.3.3 Phase II: Driver Consistency

In phase II we focus on the impact of different driver consistency requirements DC, PDC, and NODC. Since there is no indication that the order strategy significantly influences the impact, we only tested milk run configurations with heijunka levelling. In order to investigate dependencies between driver consistency and arrival time consistency, we include both a scenario with ATC and one without any time consistency requirement (see also Table 8.5).

Table 8.5: Model configurations of phase II

DC	PDC	NODC
HEI FIXSC NOTC	HEI FIXSC NOTC	HEI FIXSC NOTC
HEI FIXSC FRATC	HEI FIXSC FRATC	HEI FIXSC FRATC

Impact on Solution Quality

In cases in which the weight β of the partial driver consistent formulation is sufficiently small, this formulation eliminates symmetry mainly caused by different feasible assignments of routes to vehicles or drivers: That is, from all distance optimal solutions the one is favoured which causes a minimum number of driver inconsistencies.

A glance at the solutions of the NODC and PDC cases suggested that most of the PDC optimal solutions can be reached by reassigning tours which are part of the optimal solution in the NODC case to other vehicles. To prove that, we implemented the following post processing step for the NODC solution: We reduce the set of feasible routes to the ones which are part of the NODC solution and rerun the model adding the PDC constraints. In all 20 scenarios – we tested NOTC as well as FRATC cases – we record the same results in terms of distance and number of driver inconsistencies for the PDC case and the NODC case with a reassignment post processing step. This result suggests that a decomposed approach in practice seems not to deliver significantly worse solutions in terms of inconsistencies compared to an integrated version.

According to Figure 8.3, the total number of driver inconsistencies aggregated over all instances could be reduced from 55 to 17 for the NOTC configuration and from 50 to 18 for the FRATC case. The maximum number of inconsistencies over all instances is 3: That means that three out of eight suppliers are serviced by two different drivers.

Figure 8.4 shows that enforcing strict driver consistency (DC) leads to an average increase in driving distance of 6%, but with a peak of up to 15% for both the NOTC and FRATC cases. At the same time, over all instances the aggregated number of vehicles goes up by 5 (from 79 to 84): In six cases there is no increase at all, however, for one instance the number of vehicles increases significantly from 8 to 10. The NOTC configuration shows very similar cost increases, i.e. there is no significant effect of arrival time consistency requirements on driver consistency.

Please note that our results are in line with the ones for larger instances solved heuristically for PVRP with fixed pick up patterns: Kovacs et al. (2014), for example, report that cost can be reduced by 6.5% by allowing two drivers per customer instead of one, and Feillet et al. (2014) report on improvements of 7.5% for the instances of Groër et al. (2009) if the driver consistency requirement is relaxed.

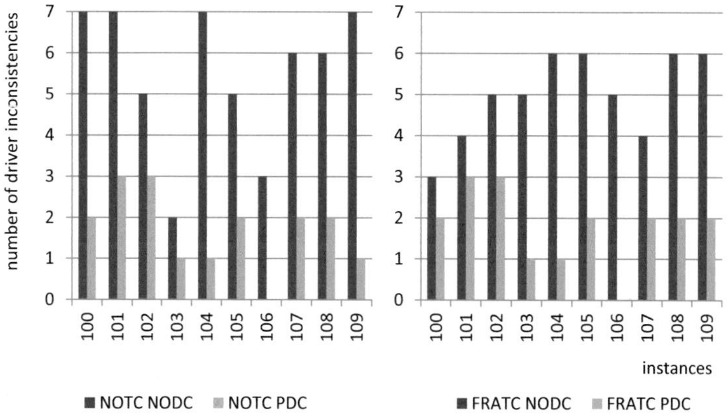

Figure 8.3: Resulting driver inconsistencies per instance for the configurations NOTC (left) and FRATC (right) for the DC and PDC formulation

The partially high increases in cost confirm the assumptions that a strict driver consistency should not be enforced if it is not necessary due to organisational reasons, and that a partial driver consistency is clearly beneficial. It might be helpful to know the cost of a total driver consistent schedule if the decision maker does not want a strictly hierarchical objective function, but tunes the parameter β in a way that a certain detour is accepted if a driver inconsistency is avoided. Anyway, in practice, it is usually hard to determine the benefit for driver consistency in monetary terms.

Besides the defined partial consistency requirement, for real-world use cases, it also might be helpful to ask for driver consistency only for a certain set of suppliers, for which the driver needs special skills or whose geographical or organisational circumstances are complex.

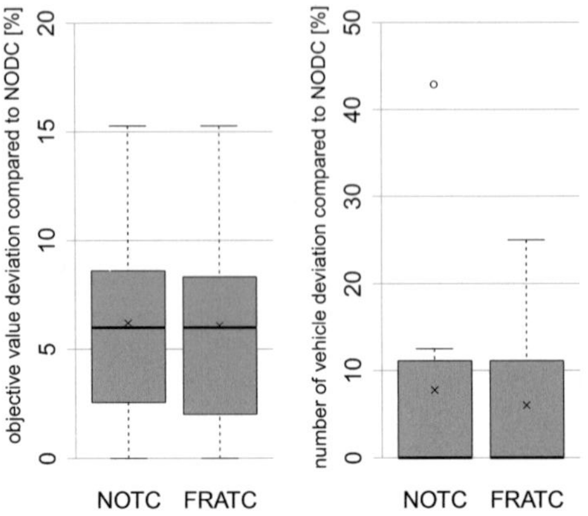

Figure 8.4: Boxplots with the mean value represented as a cross showing the percentage increase of the objective value (left) and the number of vehicles (right) if strict driver consistency is enforced assuming a fixed service choice (FIXSC)

Impact on Run Times

Considering the run times for the FRATC models, the symmetry breaking effect of DC and PDC seems to be noticeable in 4 out of 10 instances: The run times are shorter for both the PDC and the DC configuration compared to the NODC one. However, in total the average runtime of 110 seconds increases by a factor of 1.1 for the PDC and by a factor of 1.7 in the DC case. In the NOTC model, there is an increase for all instances but from a very low level: the run time for the NODC configuration is on average 0.4 seconds.

8.3.4 Phase III: Scheduling Aspects

In phase III we focus on the impact of the two different scheduling types ATC and ITC, and we analyse the effect of a fixed starting time in the arrival time consistent model (FIXATC). We furthermore check, if driver consistency and the scheduling requirements are conflicting consistency goals. However, since the schedule quality of a partial driver consistent solution behaves as a schedule without driver consistency, we do not consider the PDC case. An overview of the considered model configurations is given in Table 8.6.

Table 8.6: Model configurations of phase III

HEI FIXSC NOTC	HEI FIXSC ATC	KAN FIXSC ITC
NOTC NODC	FRATC NODC	FRITC NODC
NOTC DC	FRATC DC	FRITC DC
	FIXATC NODC	
	FIXATC DC	

The size of the dynamic time windows, obviously, has an impact on the schedule, therefore, we also consider different values for the parameter L – besides the standard value 5, we analyse the results for size 0 and size 2. The value $L = 5$ corresponds to 17% of the maximum tour duration $D = 30$. Assuming in a real-world application a tour duration restricted to 9 hours by law, this would correspond to a time window size of 1.5 hours.

Impact on Solution Quality

Arrival Time Consistency The FRATC DC and NODC schedules with a maximum time deviation of $L = 5$ show nearly the same driving distances as the corresponding schedules in which no time consistency constraints are imposed. The average increase is 0.15%, the maximum 0.94% and in 14 out of 20 scenarios the increase is even 0.

As the frequency in this configuration is fixed, the minimal possible inventory cost can be given in advance. The deviation from this inventory optimal schedule is bounded by $L = 5$, and the routing cost is less than 1% above the cost occurring in a case in which the scheduling and, hence, the inventory is ignored. Hence, this result suggests that the solution is close to an optimal one resulting from an IRP model considering inventory and routing cost simultaneously.

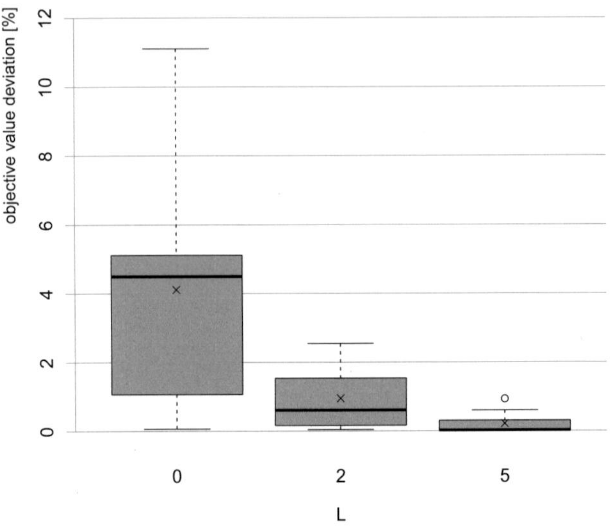

Figure 8.5: Percentage in increase of the objective value compared to the baseline model (NOTC NODC) for the FRATC NODC configuration for parameter values $L = \{0, 2, 5\}$

From Figure 8.5 can be seen that if the size of the dynamic time window is reduced to 2 – we only considered the NODC case, the average increase compared to the base line solution without time consistency is still slightly below 1%, with a maximum of 2.6%. Reducing L to 0, i.e. enforcing the exactly same arrival time for every visit at one supplier, the average increase is 4% with a maximum of 11%. That means, if the inventory costs are high enough, such that the choice of the deviation parameter L influences the routing decisions, a model explicitly balancing inventory and routing cost (as in (Coelho et al. 2012)) might be beneficial. However, in practice, it is not expected that inventory cost for milk run parts are so high that a period of the length of a time window has considerable cost effects. Moreover, in practice it is not possible to guarantee exact arrival times without including large time buffers into a tour. A decomposition of the inventory and the routing decision, therefore, seems to be acceptable since with the presented model both low inventory cost and cost efficient, regular, arrival time consistent schedules can be reached. This is especially important in a case in which the milk run cycle time is longer than a day.

If the tour start times of an arrival time consistent schedule are fixed (FIXATC) to the beginning of the period, the cost increase for our 10 instances are rather moderate if the time window size is 5. Even without allowing waiting time, the driving distance is on average only 0.8% (with a maximum of 4%) longer than in the ATC solution with a free start time for the tours. However, if the time window size is reduced, the increase is – as expected – considerable, up to 15% and with an average of 7.5% in case of $L = 0$.

The cost increase due to smaller time windows is in both cases, the FRATC and FIXATC models, a mixture of two effects: The number of tours driven over the week increases slightly (and causes longer travel times), and the number of tours in which the sequence of stops is not distance optimal due to the dynamic time window constraints increases with reduced time window sizes: For the FRATC model with $L = 5$, there are – aggregated over all instances – 3 tours part of the optimal solution which are longer than the corresponding distance optimal tour through their stops. For the same model with $L = 2$ and with $L = 0$ the number is 6. In the FIXATC the increase is from 8 to 12 to 15 respectively.

Please note that these results also correspond to the results on bigger instances for PVRP with fixed patterns: Kovacs et al. (2014) and Kovacs et al. (2014) confirm for their test instances that "high levels of time consistency can be provided with small increases in the travel cost" (Kovacs et al. 2014) – they report cost increases of 0.63% and 1.62% (see (Kovacs et al. 2014)), on average. However, the authors also confirm that low values can be only reached if vehicle departure times are allowed to vary.

Inter Arrival Time Consistency For models requiring inter arrival time consistent schedules, our time limit of 10 hours was not enough to get optimal solutions for all instances: For the configuration with driver consistency, 7 instances could be solved to optimality, for the model without driver consistency only 3. Since the optimality gaps go up to 25%, we restrict the analysis to the optimal results.

The driving distance compared to ATC schedules increases in all instances: For the driver consistent solutions distance rises by 27% on average and by 56% as a maximum. At the same time, for two instances the number of vehicles had to be increased from 2 to 3 and for one even to 4 in order to get a feasible solution, and the total number of tours over all instances increased from 56 to 76. If no driver consistency is required, the distances increase by 8% on average with a minimum of 4% and a maximum of 12%. Even if we consider only the results of three instances, it can be taken as an indication that the inter arrival time consistency interferes more strongly with driver consistency than the arrival time consistency. This is supported by the fact that even if we take into account the best found solutions for the NODC configurations, the average increase by 9% is much smaller than the increase of 28% in case of the DC model.

In Table 8.7 we show the arrival times and inter arrival times for an ATC and an ITC plan for one example instance. This is the instance in which the smallest increase (4%) in driving distance of the ITC compared to the ATC case occurred. For example, the schedule for supplier 2 in the ATC case is easy to memorise, and this regularity might quickly lead to stable and reliable processes. In contrast, in the ITC schedule the arrivals at supplier 2 take place in all three different shifts – assuming a shift length of 30 and a start time of 0 for every day.

Only for expensive items from a supplier which has the same opening hours as the plant this might be a beneficial solution. It promotes low inventories and a smooth flow of goods with a high degree of inventory visibility at both the supplier and the plant site. As before in the ATC case with $L = 0$, we should admit here that in case of constant cycle schedules, an IRP model as proposed in (Coelho et al. 2012) – explicitly considering the inventory cost and enforcing a certain level of volume consistency – might be more appropriate: In this case, a trade off between additional inventory cost resulting from the deviation of the optimal inter arrival time and the routing cost can be achieved. The routing cost increase of 14% in our experiments might not be compensated by approximately constant inter arrival times and lot sizes over the course of the week.

Table 8.7: Resulting schedules for the FRITC (upper) and FRATC (lower) configuration for one example instance (both FIXSC NODC)

	Supplier arrival times					Inter arrival times					
supp.	Mon	Tue	Wed	Thu	Fri	Mon	Tue	Wed	Thu	Fri	ideal
	FRITC					FRITC					
1	23		70			258		268			263
2	7	33	55	86		131	126	136	131		131
3	12		59			258		268			263
4	7	29	55	81		126	131	131	136		131
5			50					525			525
6		20		72			263		263		263
7	5		52			258		268			263
8	18		66			258		268			263
	FRATC					FRATC					
1			13		12			208		317	
2	6	4		4	4	103	210		105	107	
3		13		13			210		315		
4	10	8		8	8	103	210		105	107	
5					14					525	
6	19		24			215		310			
7			11		16			215		310	
8		19		19			210		315		

Impact on Run Times

Figure 8.6 gives an overview of solution times for different arrival time consistent configurations and different values of parameter L: Obviously, the run times increase with decreasing time window sizes – in case of one instance, the optimality could not be proven within 10 hours in the FRATC NODC $L = 0$ case, but the optimality gap is only around 3%.

Especially in case of the larger time window sizes, the optimal solution was often found within less than half a minute, and the rest of the run time was invested to prove optimality. The graph also shows that omitting driver consistency results in shorter run times, especially with regard to outliers.

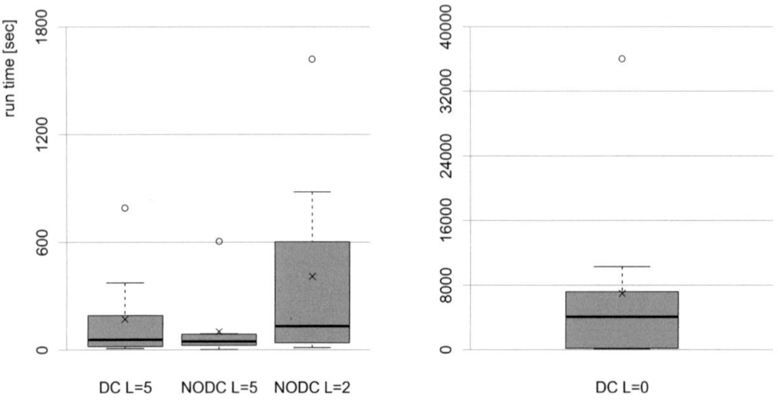

Figure 8.6: Run times for FRATC configurations with and without driver consistency and for different values of parameter L (consider different scales of y axis)

As already mentioned, the run times for inter arrival time consistent models exploded, and we could only solve 10 out of 20 instances to optimality. Obviously, the symmetry breaking effect of including driver consistency is effective, as 7 out of these 10 optimal results occurred in DC cases.

We furthermore checked the experimental results of Groër et al. (2009) with regard to the precedence principle for ATC schedules with driver consistency: The idea of the precedence principle is that, whenever a customer is served on the same tour before another one, this precedence relation is preserved for all tours on all other days. In just one out of the 10 small instances of Groër et al. (2009), this precedence principle is not fulfilled in the optimal solutions.

We observe a similar result: Only in 2 out of 10 schedules, the precedence principle is not fulfilled if arrival time consistency and driver consistency are both imposed. However, if we omit driver consistency, the principle is violated in 7 solutions. This suggests that during the search this principle should not be applied to decrease the search space as proposed in (Groër et al. 2009).

8.3.5 Phase IV: Service Choice

In the last phase we want to show the effect of allowing service choice. We test the three introduced service choice types with arrival time consistent schedules and with and without driver consistency. Due to the long run times, we excluded the ITC case and tested the scenario with free service choice without considering any scheduling constraints. We expect similar effects for the other cases (see also Table 8.8).

Table 8.8: Model configurations of phase IV

FIXSC	ALSC	FRSC
HEI NOTC NODC	HEI NOTC NODC	HEI NOTC NODC
HEI NOTC DC	HEI NOTC DC	HEI NOTC DC
HEI FRATC NODC	HEI FRATC NODC	
HEI FRATC DC	HEI FRATC DC	

Impact on Solution Quality

Since in this tests we do not explicitly consider inventory holding cost, the only incentive for choosing higher frequencies in the ALSC case is routing cost. In just one out of our 10 random instances, this effect arose: In this instance, three suppliers are visited with a frequency of three instead of two, resulting in a cost reduction of 3% – for the FRATC NODC, as well as for the NOTC NODC configuration. For the cases with driver consistency the cost reduction for the same instance is neglectable with less than 0.3%. If inventory costs are small compared to routing cost or not considered, these results seem to suggest that the reduction of the search space by forbidding service choice does not deteriorate the results significantly.

With the FRSC configuration we determine the lower bound of routing cost, since every supplier needs to be served at least once per week and the only capacity restriction considered is the one of the vehicles. As it can be expected, the number of visits performed within a schedule and the corresponding transport cost are reduced dramatically. Due to long run times in the ATC case, and, since this is only a lower bound, we only conducted the experiments for the configurations without scheduling requirements. For both the NODC and the DC case the average cost saving compared to the FIXSC is around 40%. However, the total number of visits over all instances drops from around 250 to around 110 in both configurations.

Impact on Run Times

The solution times for the ALSC version increase by an average factor of 22.5 for the FRATC DC and 51.6 for the FRATC NODC case compared to the FIXSC configurations. However, the median is 0.9 and 2.5 respectively. That means, the solution times are rather unstable. In some cases, the ALSC case is even solved faster, but we have outliers with a factor of up to 400. The run times for the FRSC case compared to the FIXSC configurations (NOTC) are considerable: On average the runtime is longer by a factor of 700 (median 460) and for the DC case by a factor of 190 (median 150).

However, in both cases more restricted frequencies, i.e. only the lowest and the second lowest, should lead to smaller run times. Furthermore, in the ALSC case results from the FIXSC model could be used as a starting solution delivering a good upper bound.

8.4 Summary

In this chapter we introduced reasonable model variants for the milk run scheduling problem based on the PVRP-SC along with the cyclic inventory models introduced in the preceding Chapter 7. In all cyclic inventory models – CCI, CDI, and CDI-LT – it is assumed that the optimal inter arrival times are approximately met in the milk run schedule for all suppliers. Therefore, we extended the PVRP-SC by constraints assuring inter arrival time or arrival time consistency. Hence, we proposed an integrated milk run tour and scheduling problem, which assigns at the same time a frequency to every supplier minimizing operational cost. We furthermore proposed driver extensions expressing driver and partial driver consistency.

In order to be able to evaluate the different possible configurations of the milk run scheduling model, we solved small instances to optimality. To this purpose, we proposed an a priori column generation similar to an approach of Andersson et al. (2011) used for another vehicle routing problem with dynamic time windows.

The evaluation was conducted on 10 artificial instances with 8 suppliers and a milk run cycle time of 5 days: The tests were split in 4 phases each dedicated to show a certain aspect of the milk run scheduling configurations.

Kanban versus Heijunka Considering inter arrival time consistent milk run schedules the objective values for unlevelled (CDI) and levelled order strategies (CDI-LT) do hardly differ in our 10 instances. That means, the quality of the resulting solutions is the same. However, in the unlevelled case the vehicle capacity needs to be considered in the path flow model, and, thus, the solution times increase. All instances for all unlevelled configurations were solved on average in 490 seconds, while the heijunka configurations could be solved in 90 seconds.

Driver Consistency Our experiments showed that strict driver consistency for all suppliers might be expensive and increases the travel cost by up to 15% with additional vehicles needed. If strict consistency is not necessary, the number of different drivers per supplier can be significantly reduced by applying the concept of partial driver consistency to the model. However, for our instances the same level can be reached by reassigning tours in a post processing step to the drivers resulting in the least number of changes. The results furthermore showed that ATC and DC do not conflict with each other.

Scheduling Aspects The sensitivity analysis of the ATC model to different sizes of the dynamic time window L showed that sensibly chosen time window sizes lead to a very low transport cost increase of less than 1% for all instances and models with and without driver consistency. Even a very small time window size still leads to moderate cost increases, which are on average still below 1%. These results suggest that the proposed decomposed IRP model is adequate for weekly recurring milk runs leading to solutions close to minimum routing and inventory cost resulting from an integrated model. The average run times of 170 seconds over all ATC configurations are as well acceptable, especially, since on average after half a minute the optimal results have been found.

In contrast, the computation times for the ITC configurations explode, and only 7 out of 20 runs could be finished proofing optimality within 10 hours. In terms of solution quality, the results showed, that ITC and DC are clearly conflicting consistency goals leading to cost increases of up to 56%. The cost increases without applying driver consistency are still on average 8% and up to 14% for a sensibly sized parameter L.

Both the run time and the quality results suggest that an integrated IRP model – considering inventory cost and enforcing a certain level of volume consistency – might be preferable for producing good quality results in terms of routing and inventory cost.

Service Choice The increase of computation times for allowing ALSC compared to FIXSC is considerable. In contrast, only in one out of our 10 instances a small decrease of routing cost could be achieved by increasing the minimum frequency by one. This suggests that forbidding service choice for the sake of faster run times might not deteriorate the results considerably. A free service choice increases the run times even more, however, it might deliver a good lower bound for routing cost.

9 Transport Concept Assignment and Milk Run Scheduling

In Chapter 4 and 6 we argued that the decisions on the transport concept assignment and the milk run scheduling – including the frequency assignment for every supplier – should be taken simultaneously in order to achieve minimum operational cost. In this chapter we propose an integrated model for the transport concept assignment and milk run scheduling decision considering, besides milk runs, the two main concepts currently applied in automotive companies: point-to-point transports and groupage services executed by an area contract freight forwarder – in the following referred to as area forwarding (AF). That means that the model considers the question, which suppliers can be combined to a milk run and which suppliers should be served through the AF networks or by point to point transports assuming deterministic demands.

As discussed in Section 2.1, this decision depends for very expensive or critical parts on thoughts concerning reliability: The regularity of milk runs results in more reliable processes for a couple of reasons. However, as discussed in Chapter 2 we assume in the following that we focus on the question for which supplier relations the milk run concept might be beneficial in terms of operational cost keeping the additional advantages in mind.

In contracts for area forwarding services, point-to-point transports are often included. That means, above a certain weight or volume limit of an order a point-to-point tariff is incurred. Such a tariff table can also be determined ex ante if separate contracts for area forwarding and point-to-point transports exist. Thus, from a modelling point-of-view we can reduce the transport concepts to AF – including point-to-point transports – and milk run services.

We use the execution of all orders by AF services as baseline solution for our experiments since it is the simplest transport concept from the perspective of the consignee: The consignee outsources both the tour planning task and the risk for inefficient tours.

Before presenting the integrated Transport Concept Assignment and Milk Run Scheduling (TC-MRS) model, we specify and discuss tariff systems for the transport concepts introduced in Section 2.1.2 in more depth. In our computational experiments we analyse the resulting plans of different instances and compare the optimal results with the ones we received from the weight based allocation heuristic known from practice.

9.1 Tariff Systems of the Considered Transport Concepts

As already discussed in Section 2.1.2, the costs for groupage services – and hence area forwarding – usually depend on the weight and the distance between the source and the destination, but not on the actual tours. In contrast, milk run costs depend on the driving time and distance and usually not or only to a small extent on the weight or volume. In the following we further detail the cost structures introduced in Chapter 2 and discuss, how good cost estimations for the milk run case can be generated.

Activity Based Costing

If milk runs are served by vehicles and staff of the producing company itself, the cost for tours can be estimated based on the historical data of the company. However, if the tours are served by a logistics provider, the cost for the company corresponds to the price that the logistics provider offers for a concrete tour. This depends on a couple of factors not known to the buying company, such as the available capacity of the logistics provider and the share of deadhead kilometres of the logistics provider within the area of the corresponding tour. However, to take the transport concept decision the buying company needs a good cost estimation for milk runs.

An activity based costing approach (ABC) provides (for an introduction see (Gudehus 2012)) a good basis for this estimation: One estimates the cost values for all relevant activities of the transport concept based on statistical information for a region or a country. The statistical sources can be public – for example, the Eurostat or the Federal Statistical Office of Germany for diesel or labour costs – or historical data from the buying company. As an example from practice, a slightly adapted version of the ABC model of the LOCOM Consulting GmbH is introduced in Table 9.1. It includes, for example, volume, distance, time, routing and the number of used vehicles as cost drivers. For a complete tour all cost drivers are known and the total cost can be easily evaluated using the estimated values.

Table 9.1: Simplified overview of the activity based costing model of the LOCOM Consulting GmbH

		Cost driver				
		volume	distance	time	routing	nv.
Transport	staff costs	–	–	✓	–	–
	diesel costs	–	✓	–	–	–
	fixed veh. costs	–	–	–	–	✓
	var. veh. costs	–	✓	–	–	–
	toll costs	–	–	–	✓	–
Hub	handling costs	✓	–	–	–	–

veh. vehicle
var. variable
nv. number of vehicles

The resulting cost for a tour need to be increased – and in some cases even decreased – in order to receive the market prices offered by a logistics provider. This is usually done by adapting the cost by a percentage value. A profit margin causes an increase. A further increase or decrease comes from the effect that most logistics costs depend on the capacity utilization of the resources within the logistics network (see (Gudehus 2012)) and, hence, are specific to the logistics provider and the region: That means, the cost depends on the market volume within the area, the market share, and the available resources of the logistics provider.

The quality of these estimations depends on the quality of the statistical data and the knowledge of region specific situations. However, in practice the estimations based on the activity based costing approach seem to be a sensible planning criteria for the next planning period. They can be further improved by establishing a system for logistic costs controlling, which compares the estimated costs with the realised costs ex post. This allows for corrections for future planning periods (cmp. (Gudehus 2012)).

It is furthermore possible that the consignee arranges fixed tariffs with a freight forwarder for regular milk runs in a long running contract of one or two years with a payment on route basis (see Section 2.1.2).

Typical AF Transport Tariffs

As described in Section 2.1.2, in an AF network the prices for a concrete order can be looked up in tariff tables based on different cost structures. Figure 9.1 gives an example for an S2 cost structure in the form of a matrix with distance and weight classes in the columns and lines and the cost per shipment [Euro] as the matrix value. The example shows the typical behaviour of transport tariffs: The cost function is degressive in volume and distance, that means, with increasing weight and increasing distance the cost per unit decreases. It is a typical example for tariff tables of area forwarding systems.

Table 9.2: Common pricing criteria for transport price tables related to the categories distance, volume and transshipments

Distance	Volume	Transshipment
distance classes for a region [km]	pallet space classes [pallet space]	handling cost
distance classes for sink (source) location [km]	loading metres [metres]	handling cost for specific transhipment points
from source location to zipcode area or from zipcode area to sink location	weight classes [tons]	
from zipcode area to zipcode area		

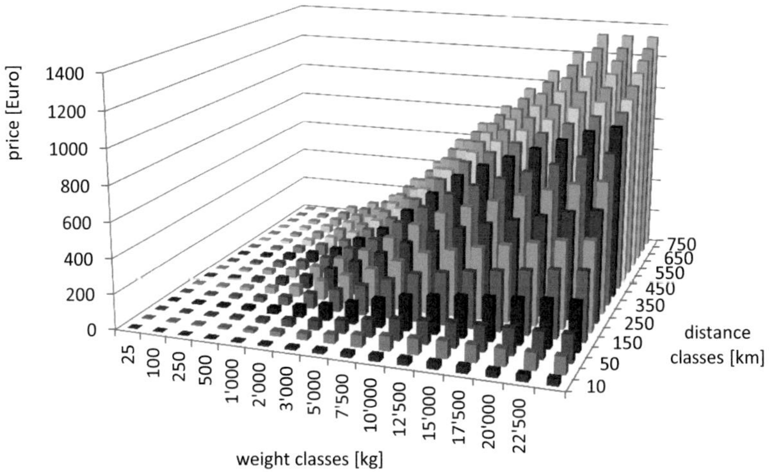

Figure 9.1: Example for a tariff table common in area forwarding networks differentiated by distance and weight classes

If the weight capacity is not the restricting dimension, the weight classes might be substituted by volume, pallet, or loading metre classes. In practice, there also exist contracts in which different price matrices are valid for different categories of products for the same region. Instead of distance classes, there might also exist specific tariffs for a source or sink location since the infrastructure around this site influences the cost. In other cases, instead of distance classes, the costs depend on the zipcode areas/countries to which or from which the shipment is delivered. Often, handling costs for transshipments within a hub are also explicitly considered. An overview of some standard cases of the presented types of matrices is given in Table 9.2.

An AF contract regulating the transport prices within an area is usually valid for a period of about 2 years. Since diesel costs and toll cost are not under control of the logistics provider and, at the same time, have a high impact on the total costs, the tariffs often have a variable component based on current diesel price and toll cost. Based on this, the matrix is, for example, adapted every three months.

However, irrespective of the cost structure and the considered dimensions, for the following we can assume that the transport cost for an order sent through the groupage service network can be determined based on existing AF contracts in advance. Remember from Chapter 6 that we consider the simple – but not unusual – case in which one tariff table for the pre- and the main-leg is arranged.

9.2 Integrated Transport Concept and Milk Run Scheduling Model

As discussed in the preceding section, in a typical planning situation the prices for the AF network are regulated in a running contract, while the milk run costs are estimated by means of an ABC approach. With these data as an input, the extensions of the MRS base model to an integrated model for assigning suppliers to transport concepts and building the milk run schedules at once are straight forward: We introduce a new set of binary decision variables y_i indicating whether supplier i is assigned to the AF network ($y_i = 1$) or is served by a milk run ($y_i = 0$). The costs for a milk run – represented by a index t in the model – are evaluated by using the estimated costs and are represented by c_t. Furthermore, the parameter c_i^{AF} stands for the costs of a shipment which has the size of the weekly demand divided by the minimum frequency.

$$\text{(TC-MRS)} \qquad \min \sum_{t \in \mathcal{R}} \sum_{d \in \mathcal{T}} \sum_{k \in \mathcal{K}} c_t \cdot z_{tdk} + \sum_{i \in \mathcal{V}^c} f_i \cdot c_i^{AF} \cdot y_i \tag{9.1}$$

subject to

$$y_i + \sum_{r \in \mathcal{C}_i} u_{ir} = 1 \quad i \in \mathcal{V}^c \tag{9.2}$$

$$\sum_{t \in \mathcal{R}} z_{tdk} \leq 1 \quad d \in \mathcal{T}, k \in \mathcal{K} \tag{9.3}$$

$$\sum_{r \in \mathcal{C}_i} u_{ir} \cdot \alpha_{rd} - \sum_{t \in \mathcal{R}} \sum_{k \in \mathcal{K}} z_{tdk} \cdot a_{ti} = 0 \quad i \in \mathcal{V}^c, d \in \mathcal{T} \tag{9.4}$$

$$z_{tdk} \in \{0,1\} \quad t \in \mathcal{R}, d \in \mathcal{T}, k \in \mathcal{K} \tag{9.5}$$

$$u_{ir} \in \{0,1\} \quad i \in \mathcal{V}^c, r \in \mathcal{C}_i \tag{9.6}$$

$$y_i \in \{0,1\} \quad i \in \mathcal{V}^c \tag{9.7}$$

The TC-MRS model is very similar to the MRS base model. The only changes emerge in line (9.1) and (9.2): The objective function is extended by the cost accrued if a supplier is assigned to the set of suppliers served by the area forwarding system. Please note that we omitted the inventory holding costs since, as in Chapter 8, we do not consider them during the computational experiments. In line (9.2) we make sure that a supplier is either assigned to the area forwarding system or to a feasible schedule r. If a supplier is assigned to the area forwarding system, we do not need to assign a pattern since AF costs are independent of the days.

All extensions of the base MRS model introduced in Chapter 8 can be applied in the same way to the TC-MRS model. For the ALSC model configuration we take AF cost accrued for transports of the minimum frequency since, in case of degressive cost structures, there is no reason to increase the frequency. Remember that the minimum frequency is calculated by taking the maximum of the economic minimum frequency for groupage services and the minimum frequency resulting from capacity restrictions.

The a priori route generation procedure introduced in Chapter 8 can also be applied without changes.

9.3 Computational Experiments

9.3.1 Instances

For testing the impact of the TC extension to the MRS model we use the same instances as introduced in Section 8.3.1. As before, the cost for the milk run tours only include variable cost directly corresponding to the Euclidean distances. The extension to other cost structures is simple and is considered in the use case of Chapter 10. Here we keep the milk run cost fixed and vary c_i^{AF} to show the impact of different tariff levels. In order to reach consistent cost values, we define c_i^{AF} proportional to the cost of serving the stop i with an exclusive milk run. That means, we simply multiply the travel cost for travelling from the plant to i and back to the plant – corresponding to a round tour – with the parameter $p = \{0.3, 0.4, 0.45, 0.5, 0.55, 0.6, 0.7, 0.8, 0.9, 1\}$. Please note that we do not assume degressive tariffs for the volume and distance since this feature only affects the preprocessing step of economic frequency assignment, but not the transport assignment decision. For our instances we set the frequencies randomly (see Section 8.3.1).

9.3.2 Phase I: TC-MRS with Different Extensions

The main focus of this evaluation is to check how the introduced model variants react to varying tariff levels. Due to long run times we leave out the model configurations assuring inter arrival time consistency. Furthermore, we do not consider the free service choice since we assume that a minimum frequency must be accomplished in case of both transport concepts. Partial driver consistency is not considered since we assume that good results can be achieved by applying the post processing step proposed in Section 8.3.3. Consequently, we consider the configurations listed in Table 9.3.

Table 9.3: Model configurations for testing the TC-MRS

FIXSC	ALSC
HEI NOTC NODC	HEI NOTC NODC
HEI NOTC DC	HEI NOTC DC
HEI FRATC NODC	HEI FRATC NODC
HEI FRATC DC	HEI FRATC DC

Impact on Solution Quality

Comparing the FIXSC and the ALSC configurations of the TC-MRS reconfirms the observations described in Section 8.3.5: The additional degree of freedom resulting from the possibility to assign a higher frequency does not lead to significantly better solutions. It is just one instance for which the ALSC solution yields on average over all configurations 1% lower cost. Hence, in the following we focus on the FIXSC case.

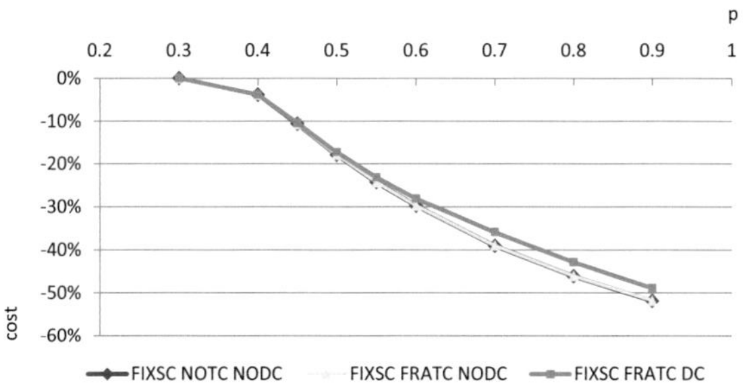

Figure 9.2: Average TC-MRS cost over all instances compared to the base case of exclusive AF shipments for three different model configurations differentiated by the AF cost parameter p

The cost of the base case scenario – an exclusive usage of AF services – are used as a baseline for Figure 9.2: Depending on the AF cost parameter p the savings resulting from considering milk runs as an alternative transport concept vary between 0% and 50% irrespective of the model configuration. That is, in case of $p = 0.9$ the exclusive AF scenario is twice as expensive. However, in this scenario the AF cost are extremely high.

Figure 9.3 allows a more detailed analysis of the results: The upper graph shows the share of AF visits, that is, the number of AF visits divided by all visits, both aggregated over all 10 instances. The lower graph shows the share of AF costs, that means, the AF costs divided by the total cost corresponding to the objective function value, also aggregated over all 10 instances. The values on the x-axis represent the cost parameter p. For the smallest value $p = 0.3$ all visits are served through the area forwarding system and, hence, no milk run costs occur. The more expensive the area forwarding system becomes, the more visits are served by a milk run.

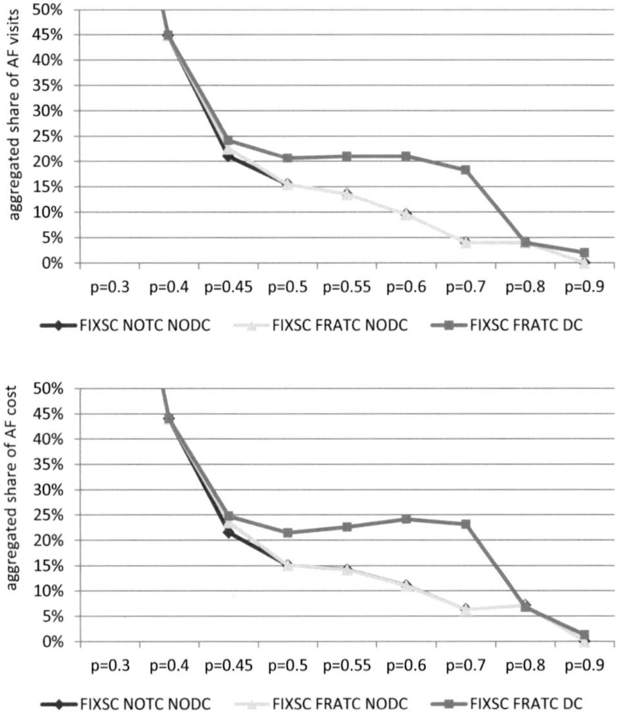

Figure 9.3: Share of AF visits and AF cost aggregated over all instances for three different model configurations differentiated by the AF cost parameter p

The figures furthermore show that the arrival time consistent schedules FIXSC FRATC NODC contain nearly the same number of AF visits and result in nearly the same AF cost as the baseline schedules without consistency requirements FIXSC NOTC NODC. That means asking for arrival time consistency – with $L = 5$ – does not have the effect of shifting more visits to the area forwarding network. This corresponds to the observation in Section 8.3.4, that ATC with a reasonably sized parameter L is not expensive. For the AF orders, delivery profiles along with time windows of size L can be arranged with suppliers and the area forwarding provider. If this is done, the AF schedules are also arrival time consistent.

As it can be expected from the results of Section 8.3.3, asking for driver consistency (FIXSC FRATC DC) leads to a shift of visits to the area forwarding system compared to a solution without driver consistency requirement. Anyway, it is questionable if it makes sense to require driver consistent schedules at the cost of serving more stops through the area forwarding system, in which the person serving the stops can not be influenced at all. As before, it rather seems to be beneficial to reassign the tours in a post optimization step to reach partial driver consistency

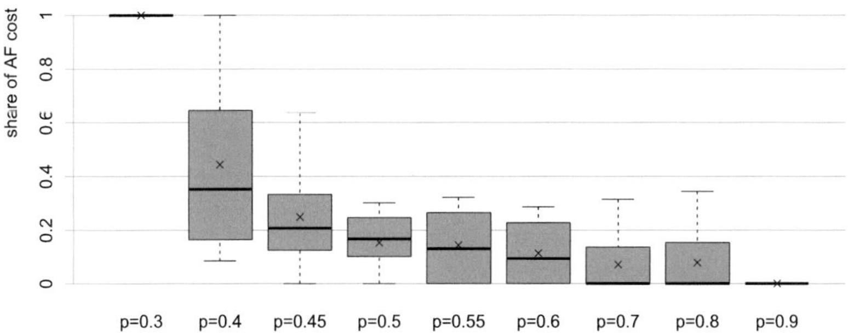

Figure 9.4: Boxplots with the mean value represented as a cross showing the share of AF costs over all instances for different values of p in case of the FRATC NODC variant

If we compare the share of AF cost on a more detailed level – as in Figure 9.4 – it becomes obvious that the transport concept decision strongly depends on the instance, that means, on the concrete geographic situation and the shipping volumes. In case of the AF cost parameter $p = 0.4$, there exists an instance in which still all visits are assigned to the area forwarding network, while in other instances up to 90% of the visits are served by arrival time consistent milk run schedules. In contrast, in scenario $p = 0.8$ there exists an instance in which still 10% of the visits are served through the AF network, which causes 34% of the total cost. However, this is still cheaper than serving these visits by a milk run.

In Figure 9.5 we show the number of AF and milk run visits and the resulting costs for two rather extreme instances. Instance 104 has a beneficial structure for milk runs, while in instance 105 visits are assigned to the AF network even in the $p = 0.8$ scenario. Remember that this means that the AF network cost reaches 80% of the cost of an exclusive milk run service. On average over all instances, the relationship between the number of AF visits and the corresponding costs are linear. That means the share of the number of AF visits corresponds to the share of AF costs. However, in individual cases, the relationship is not linear as one can see in case of instance 105 in Figure 9.5.

The relationship between the AF cost parameter p and the average capacity usage over all instances is shown in Figure 9.6: The time capacity usage is calculated by dividing the total tour time by the maximum allowed tour duration D^T. The volume capacity usage is defined by the maximum load reached at the last stop before the plant divided by the capacity Q. As expected, both measures decrease with the increase of the AF cost parameter p since gradually more and more unfavourable visits – in terms of distance and volume – need to be integrated into milk runs and the number of bad tours increases.

This relationship illustrates the fact that the risk for "bad tours" is just outsourced to the area forwarding service provider. This only makes sense if the service provider has a big number of third party clients in the area in order to be able to build good tours.

9.3.3 Phase II: TC-MRS vs. WA Heuristic

As described in Chapter 4, the assignment of the transport concept based on simple rules – for example an allocation heuristic based on the weight (WA) – is common in practice. In order to compare the results from such a simple rule based approach to the optimal solutions of the TC-MRS model, we fix the decision variables y_i to 1 for all visits i with a pick up volume d_i^{HEI} smaller than the threshold w_{min} and to 0 for all others.

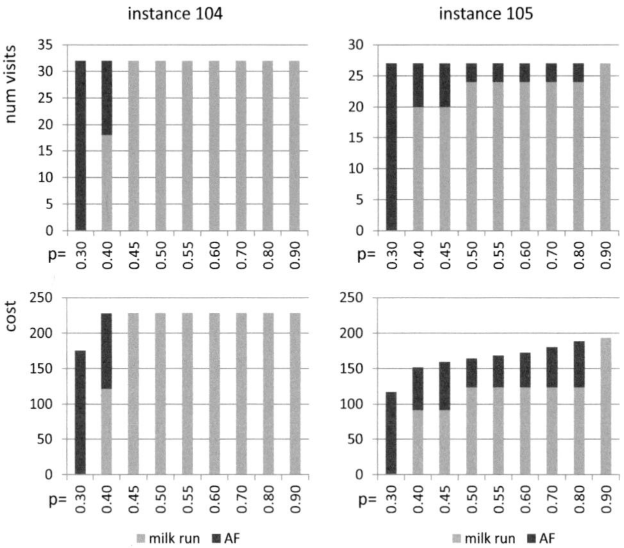

Figure 9.5: Comparison of resulting cost and visits for two instances for the FRATC NODC variant

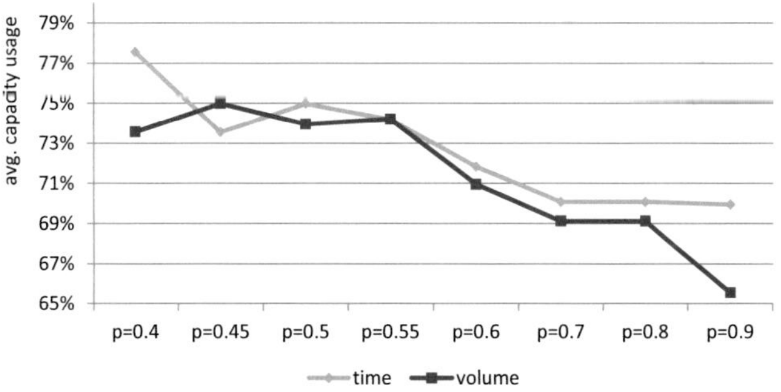

Figure 9.6: Average capacity usage of the resulting milk runs over all instances for different values of p in case of the FRATC NODC variant

Solving the resulting model means that we select the optimal milk runs for all visits with pick up volumes bigger than w_{min}. We applied this heuristic to our 10 instances with a threshold w_{min} set to 10% of the vehicle capacity Q. From the model configurations we used the FIXSC FRATC NODC to generate regular milk run schedules since ALSC did not provide any advantage in case of our instances and driver consistency should not be applied if the transport concept is not fixed.

Impact on Solution Quality

Figure 9.7 shows over all instances the percentage value of the cost increase of WA heuristic results compared to the version in which the concept decision is left to the TC-MRS model. The increase in cost is significant for all tariff levels p with an average of 35%, a maximum of 90% and a minimum of 4%. The maximum increase for all instances occurs in the two extreme cases of $p = 0.3$ and $p = 0.9$: Remember that in these cases the optimal transport concept for all suppliers was either AF or milk runs respectively. Therefore, the weight based allocation always using both concepts is very misleading.

The deviation of the cost increase is very big for different instances: The smallest average cost increase over all tariff levels is just 8% (with min of 4% and max of 31%) for one instance, while the maximum average increase is 60% (with min of 45% and max of 90%). Hence, the quality of the heuristic is very sensitive to the structure of the instances in terms of the geographic situation and the pick up volumes.

However, we must remark that the significance of these results is limited and must be transferred to bigger instances cautiously: The bigger the instances, the higher is the possibility to build cost efficient tours with the suppliers assigned to the milk run concept. In these cases, the cost increase compared to a TC-MRS solution can be expected to be smaller. Furthermore, the threshold w_{min} parameter influences the cost as well. In order to generate more reliable results, the same experiments are introduced based on the bigger use case instances in Chapter 10.

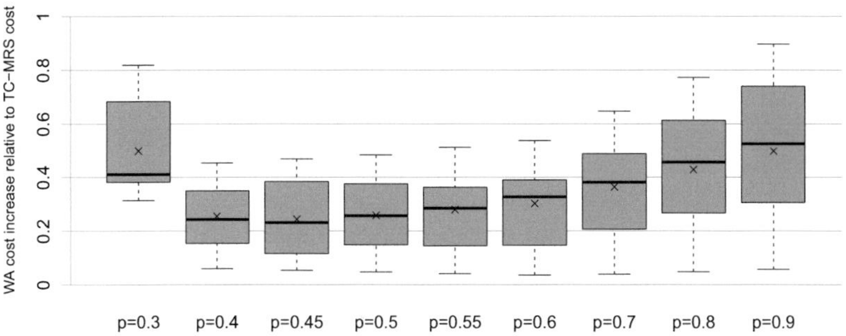

Figure 9.7: Boxplots with the mean value represented as a cross showing the cost increase of WA cost compared to the TC-MRS costs as share of of the TC-MRS costs over all instances for different values of p in case of the FRATC NODC variant

9.3.4 Impact on Run Times

Figure 9.8 shows average run times of the TC-MRS model for the different values of p. It becomes apparent that in all configurations the run times for scenarios with cheaper AF tariffs are shorter than the ones with higher tariff levels. In case of these scenarios the average run times of the TC-MRS are also shorter than the ones of the pure MRS models, even if we have more variables in case of the TC-MRS. This suggests that the cheap AF tariffs for some suppliers dominate all milk run tours containing these stops. Hence, the – up to five – visits of these suppliers do not need to be considered for the milk run scheduling.

Generally we can state that the extension to include the transport concept decision in the MRS model is very natural in case of the path flow formulation with a very small number of additional binary variables. The analysis shows that for realistic scenarios of p the run times are even shorter than in the pure MRS case.

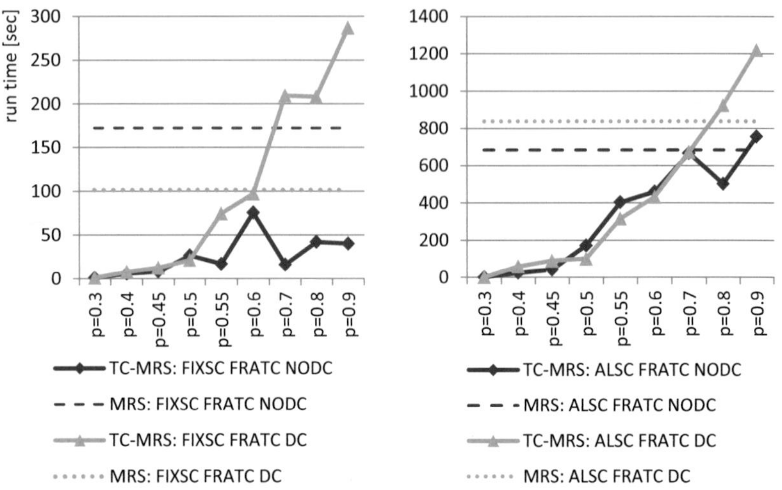

Figure 9.8: Average run times in seconds over all instances for the TC-MRS for different values of p in case of the FRATC NODC and FRATC DC variant compared to the average run time of the corresponding pure MRS model for FIXSC (left) and for ALSC (right)

9.4 Summary

We can conclude this chapter with the insight that the transport concept decision depends on a lot of different factors: The order structure with frequencies, volumes and geographic situations, the price structure for AF services, and the cost for milk runs. The results showed that it is hard to derive simple rules for the transport concept decision. At the same time, depending on the share of cost between AF and milk runs, significant cost savings can be achieved. Hence, the application of OR methods for supporting this decision seems beneficial.

10 Milk Run Design at Bosch Automotive

The computational experiments and promising results of the preceding chapters are all based on artificial, small instances. In order to check if the insights also hold for a real life example, in this chapter, we introduce a case study for a production site of Bosch Automotive in Homburg, Germany. At this factory Bosch produces parts for diesel systems with around 4.600 employees. In 2014 the company planned to produce around 8 million common rail injectors for passenger cars, around 4 million for commercial vehicles, around 2 million rails, and around 300.000 unit and in-line pumps[1].

The case study is based on historical AF shipment information of Bosch Homburg and latest vehicle cost information from LOCOM Consulting GmbH. The goal of this case study is to show a structured way to identify the potentials of milk runs as an alternative transport concept using relatively easy-to-gather data.

We start this chapter by introducing the problem setting and the available data base. We continue describing analysis steps for deriving input data for the Milk Run Design decision. We expect that this procedure is easily applicable to other cases since we use historic transport data, which are usually available due to billing reasons. We introduce the test scenarios and different model variants and analyse the corresponding results. We furthermore describe the performance – in terms of run times – and the limits of the a priori column generation approach introduced in Section 8.2. We propose an easy to calculate parameter, which can be used to get a first insight if milk runs might be a cost efficient alternative to area forwarding shipping. We conclude the chapter by discussing the effects of a decision support system for the Milk Run Design (MRD) problem on the contractual situation between the client and the logistics service provider.

[1] Source: Data for the year 2014 provided by Bosch.

10.1 Problem Setting at Bosch Automotive

Initial Situation and Goal of Case Study

At the production site of Bosch Homburg, 70% of the inbound shipments are smaller than 0.5 tons. For such a shipment structure, milk runs are typically not considered as being beneficial. Hence, all transports from German suppliers are shipped through the network of two logistics providers, which are the current area forwarding contractual partners of Bosch. Since full truck load shipments with a weight greater than 15 tons represent less than 1% of the shipments, no special FTL contract exists. Full truck loads are sent in the highest weight class through the AF network.

The standard (transport) lead time within the AF networks of Bosch are three days for the whole of Germany: On day A the material planners order parts from the supplier. On day B the parts are picked up by the logistics provider, and on day C the items are delivered to Bosch.

Based on this initial situation, the goal of this case study is to analyse if and under which circumstances milk runs are a beneficial transport concept. If savings can be reached compared to the AF base case, they need to be considered as an upper bound since we make certain assumptions on surrounding processes. However, the upper bound enables the decision maker to judge if the efforts to operate a milk run system pay off in short-term. This is especially important since the positive medium-term effects for the company, for the suppliers, and the logistics providers by reducing the variability within the transport network are more difficult to measure and might not pay off immediately.

Data Base

Historical Transport Data As already mentioned, the case study is based on historical transport data from 2013 and 2014 comprising 22 complete weeks of one logistics provider and 17 complete weeks of the other provider. For every transport from a supplier to the plant the following information is available: supplier address, plant address, pick up date, delivery date, actual weight of the shipment, chargeable weight of the shipment, and the transport cost for the shipment operated by the respective area forwarder. The chargeable weight is the maximum of the actual weight and a theoretical volumetric weight. The way how the volumetric weight is calculated is arranged in the area forwarding contract in order to avoid unprofitable shipments for the freight forwarder due to low density shipments taking up a lot of space in the vehicle in proportion to their actual weight. The actual volume – measured in m^3 or standard pallets – is not available.

Since these data are relevant for billing and accounting, they are usually easily accessible. Other data – such as the actual volume or which orders have been transported on the same vehicles – are not relevant for billing and, hence, very often this information is not stored for ex post evaluations.

In Section 10.2 we describe how we analysed the data with the goal to derive input data for the MRD models.

Vehicle Related Parameters As described in Chapter 9, the cost for a possible milk run should be estimated based on an activity based costing model. In this case study we follow the ABC model of the LOCOM Consulting GmbH introduced in Table 9.1. The LOCOM Consulting GmbH also provided us with the according cost parameters based on statistical data from official sources and the experience from many consulting projects in the German automotive industry. Besides the cost, the capacity information is needed as an input parameter for the MRD models. The following parameters are an example for a typical 40-ton truck:

Table 10.1: Vehicle parameters

Vehicle class	Capacity [ton]	Cost per km [EUR]	Cost per hour [EUR]	Mark up
40-ton truck	20	0.6100	34.7380	15.5%

The mark up is 15.5% and is calculated as follows: There is a 5% surcharge for administration and risk, and the basis of this another 10% are added as margin. The value for risk is reduced compared to the case of ad hoc tours since for milk runs the risk for the logistics provider is lower due to a higher probability to gain return freight. In the best case, the logistics provider can acquire a regular return freight as a counterpart of the milk run.

Driver Related Parameters According to the regulations of the European Union, the daily driving time shall not exceed nine hours. It might be extended to 10 hours at most twice per week (see (European Parliament and Council 2006)). However, since we solve a tactical planning problem, the maximal tour duration is assumed to be 9 hours.

Distance And Driving Time Matrix For the tour generation phase we need distances and driving times between all nodes of the network, that means, between all suppliers and the receiving plant. The geocoding and the calculation of these distances and driving times for trucks were undertaken for this case study by the corresponding commercial components of the PTV xSERVER family[2].

10.2 Transport Data Analysis

This section describes the analysis steps – and the corresponding results – executed on the transport data in order to get reasonable input for the MRD problem. Since the structure of transport data might look slightly different for other companies, it cannot be seen as a standard procedure. However, it can be understood as an analysis process model for cases with a similar data structure and a similar initial situation.

[2] For detailed information see http://xserver.ptvgroup.com/en-uk/products/ptv-xserver/ (14.02.2015)

10.2.1 Data Analysis Steps

For a better understanding we shortly introduce the analysis steps as a whole and discuss the results for the Bosch case study in the next section. The analysis can be divided in four parts:

1. Typical Preprocessing / Data Cleansing tasks, such as merging supplier IDs in case of misspellings etc. or identifying and deleting incomplete weeks
2. Building supplier categories by frequency per week and average shipping weight
3. Illustrating the shipment structure by contrasting, for example, share of cost with frequency category or share of cost with weight category and the geographic situation
4. Variability analysis of frequencies per week and cumulated weight per week

Besides typical data preprocessing tasks a problem specific process is to identify and delete incomplete weeks: This means that we need to check if the calendar weeks are fully – from Monday till Friday – contained in the records either based on the pick up date calendar week or the delivery date calendar week. Since the lead time is fixed, it does not matter on which date the analysis is based. If only a few days of the week are contained in the data set, the average frequency per week and the aggregated weight per week are biased. This also applies for weeks in which the plant is closed. If the data record period is long enough, the data sets for these weeks should be deleted.

After data cleansing, every supplier is categorised by its average shipping frequency per week and the average shipping weight using the categories of Table 10.2. The bounds for the frequency classes are all adapted by the factor 0.9 so that a supplier is classified as daily if its average frequency is greater than 4.5, and it is categorised as weekly if the average lies in the interval $[0.9, 4.5)$. This is to account for the fact that the records contain, for example, weeks with a holiday at the supplier or plant site, which results in slightly lower average frequencies.

Table 10.2: Supplier categories

<= 0.1 tons
<= 0.25 tons
<= 0.5 tons
<= 1 tons
<= 2.5 tons
<= 5 tons
> 5 tons

(a) Weight Classes

daily
weekly
2-weekly
monthly
rare

(b) Frequency Classes

The last two steps are executed in order to analyse if the milk run concept is applicable in the case under consideration: Of special interest is the comparison of the share of cost and the share of daily and weekly suppliers. If the share of cost is high for the daily and weekly suppliers, milk runs might be a good alternative to the area forwarding system. However, if a big share of cost is accounted for monthly or even rare suppliers, the savings potential of milk runs is small since no regular flow of goods exists on the inbound side.

We furthermore inspect two dimensions of possible variability, namely the frequencies per week and the cumulated weight per week, for every supplier using the coefficient of variation as measure. If μ denotes the mean and σ the variance of a parameter, its coefficient of variation (CV) is given as (see for example (Hopp and Spearman 2001a)):

$$CV = \frac{\sigma}{\mu}$$

A low variability in both dimensions means that the current production and the order policy already leads to well levelled flows of goods on the respective relation. In this case the introduction of an order policy matching the milk run requirements should be possible and cause little adaptation effort. The underlying assumption for using the weekly accumulated weight instead of the single historic shipment sizes is that equally sized shipments over the course of one week do not significantly influence inventory levels and the corresponding cost. However, if the variabilities are high, this might be the case. In case of an advanced implementation level of lean manufacturing principles in a plant, the variability measures should be low in both dimensions – frequency and volume.

10.2.2 Data Analysis Results

In this section we briefly introduce basic information related to the transport data of our case study and then mainly discuss the results of the analysis steps 3 and 4. The analysis was implemented using the free programming language and software environment R.[3]

In total we have about 4000 transport records of the considered plant available with either 17 or 22 complete weeks. The data set contains around 250 suppliers geographically scattered over the whole of Germany but with a focus on sites in the South Western part, which can be reached within 9 hours of truck driving time from the plant (see Figure 10.1).

Please note that we use the term supplier, supplier-plant-relation – or only relation – and order in a similar way. An order always stands for the input of the MRD models for one specific supplier or relation.

Shipment Structure

Weight Structure The average weight over all transport data is around 700 kg. As already mentioned, the chargeable weight is adapted whenever the density of a shipment is low. However, this applies only in 5% of all shipments. Furthermore, the average chargeable weight is less than 7 kg higher than the actual weight. Hence, for this case study it is acceptable to only consider weight restrictions in the tour generation phase.

[3] For more details see http://www.r-project.org/ (14.02.2015)

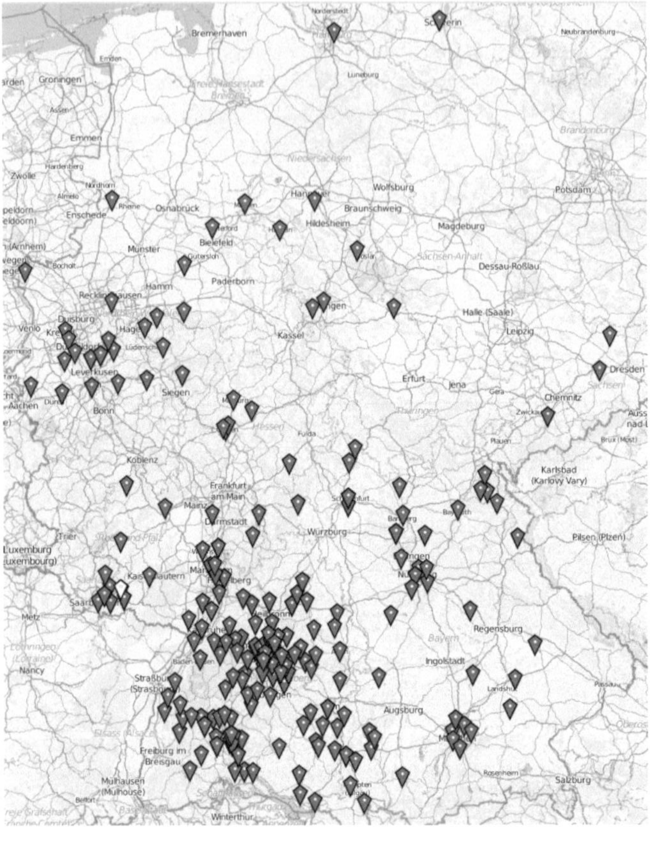

Figure 10.1: German suppliers marked in dark grey and the receiving plant marked in white (Map data © OpenStreetMap contributors)

Figure 10.2 represents the weight structure of the data set: The bars on the left hand side stand for the share of suppliers, the share of aggregated cost, and the share of aggregated weight differentiated by the weight classes introduced in Table 10.2. On the right hand side, the same figures are given accumulated. The chart shows that 40% of the supplier relations are part of the smallest weight class and, at the same time, these relations only make 2% of the total weight shipped. Up to 87% of the suppliers have an average shipment size smaller than half a ton. The other way around just 1% of the suppliers in the biggest weight class ship nearly 40% of the total weight.

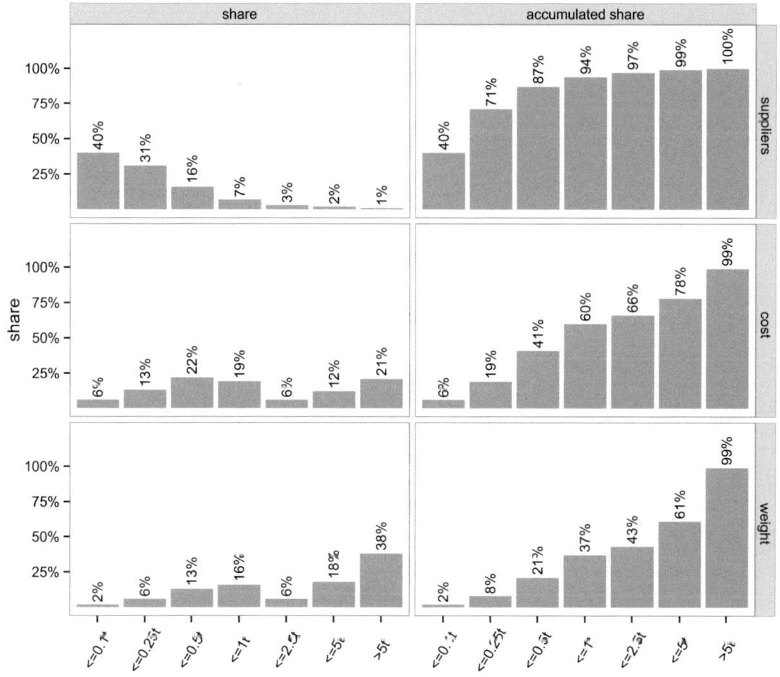

Figure 10.2: Share of suppliers versus share of aggregated weight and share of aggregated cost by weight classes

Furthermore, the chart shows that the smallest weight class contains most suppliers, but at the same time, has the smallest contribution in terms of cost and, hence, has a low potential for improvements. On the other hand, the biggest weight class containing only 1% of the suppliers represents 21% of the total cost share. From these figures it cannot be derived that milk runs are a promising alternative since many shipment sizes are rather small. However, to assess the potential of milk runs, the frequency structure needs to be taken into account.

Frequency Structure Figure 10.3 illustrates the frequency structure: The bars of the left hand chart represent the share of suppliers, the share of aggregated cost, and the share of shipments differentiated by the five frequency classes of Table 10.2. Again, the right hand chart shows the same values accumulated. The figure clarifies that just 4% of the supplier relations are served at least once per day, but another 18% are served at least once per week. On these relations 81% of the shipments cause 89% of the weight and 85% of the transport cost.

The map of Figure 10.4 shows the geographical distribution of the 54 suppliers of the class daily and weekly. Apart from the two ones marked in blue, they all can be reached within nine hours of driving time and, hence, are candidates for direct milk runs.

There exist only 2 supplier relations which are served on average twice or more often per day. Hence, daily recurring milk runs with vehicles cycling more than once a day between suppliers and the Bosch plant – as in a typical Toyota milk run system described in Chapter 3 – are not of interest. However, as expected for the German automotive industry, the daily and weekly shipments cause 85% of the cost: A milk run system with a system cycle time of one week might be a promising alternative to the AF system to reduce the transport cost.

The boxplots of Figure 10.5 show the average weight of the supplier relations classified by frequency classes. For reasons of a better readability the outliers – partly up to 9 tons – are not shown in the chart. From this chart one learns that shipment sizes of daily and weekly supplier relations tend to be bigger than for the other frequency classes. However, the median and mean value below 1 ton do not indicate immediately that milk runs served by a big vehicle with a capacity of up to 20 tons can be operated efficiently. On average one would need to serve up to 30 suppliers on one milk run to fill up the vehicle capacity wise. In such a tour, the maximum working time of a driver would be used up by loading. Furthermore, practitioners often accept no more than five to eight stops on a milk run in order to keep the severity of delays or other failures limited.

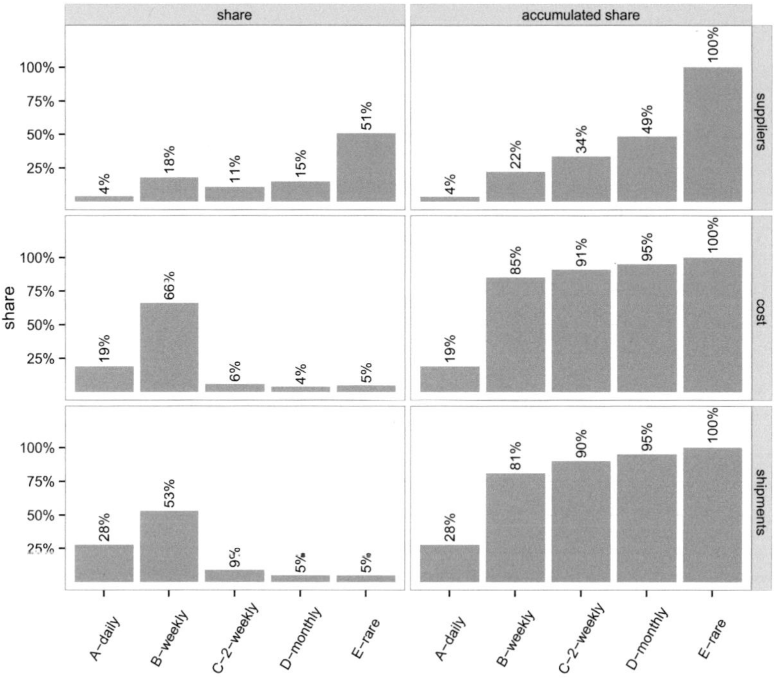

Figure 10.3: Share of suppliers versus share of shipments and share of cost by frequency classes

However, since there exist outliers in terms of weight, it is still worth to test if milk runs for supplier relations of a higher weight class – filled up by suppliers with a lower weight class – are beneficial. Furthermore, the data suggest that it might be worth to consider cheaper vehicles with a smaller weight capacity.

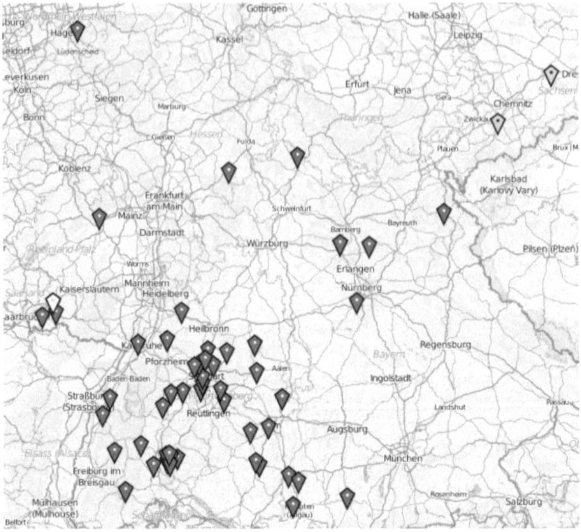

Figure 10.4: Suppliers of class daily and weekly marked in dark grey if they are reachable within 9 hours driving time and in light grey if not and the receiving plant marked in white (Map data © OpenStreetMap contributors)

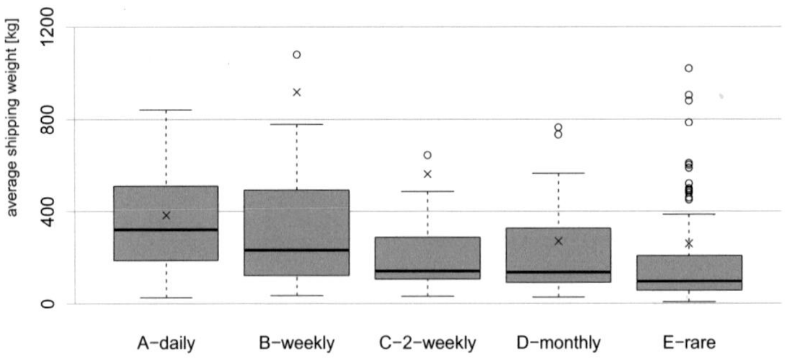

Figure 10.5: Boxplots of average shipping weight by frequency classes with mean value represented as a cross

Variability

As already mentioned, the variability analysis of the frequency and the average weight per week for every supplier gives a hint whether the adaptation of the material planning and order policy is difficult. The chart of Figure 10.6 shows the share of daily and weekly supplier relations by its coefficient of variation classes: 72% of all relations belong to classes of low variability in both dimensions, while the rest is part of the classes of a moderate variability (see Hopp and Spearman (2001a), who classify a variability as low if the value is below 0.75, as moderate if it lies between 0.75 and 1.33, and as high if it is bigger than 1.33).

This result shows that – based on the weekly demand of all parts delivered by a supplier – the variability of the weight is already quite moderate and a levelling over the week seems to be possible. One needs to keep in mind that for transportation purposes this levelling does not need to be reached on part level, but on the total shipping volume between a supplier and the plant.

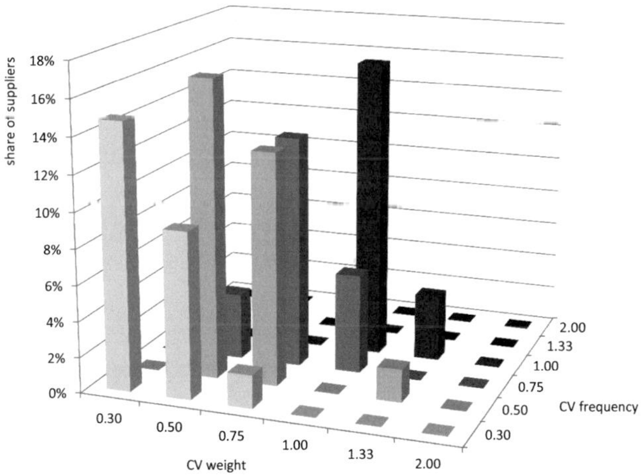

Figure 10.6: Share of daily and weekly suppliers by CV class for the aggregated weight per week and the frequency per week

However, as already mentioned, it is necessary to check for every supplier relation if the order policy can be changed to regular shipment profiles for example by applying the delivery profile assignment approaches of Schöneberg et al. (2010) or Meyer et al. (2011) introduced in Section 5.2.1. In practice there exist several reasons why the shipment profiles do not result in balanced lot sizes: The main obstacle is usually an unbalanced production in terms of supply parts so that a balancing of shipment sizes over the scheduled shipments – which can be also week spanning – is not possible or results in high inventory cost. Another reason might be the prolonged lead times: If a supplier-plant relation is only serviced once a week, the order is shipped up to three days earlier than in a case without fixed delivery profiles. This might not be possible because the concerned parts are expensive, have a high risk of obsolescence or the concrete configuration for a part is only known three days in advance. However, for all suppliers with a shipping frequency bigger than two the lead times of a milk run system are not longer than in an AF system.

To apply the delivery profile approaches above, much more detailed information is necessary: The raw demand resulting from production planning needs to be calculated. Detailed packing data are needed in order to be able to determine shipping weights and volumes from the raw demand. Inventory cost needs to be determined in order to assess the order policies cost wise. Since these data – even in big companies – are difficult to gather, we propose to first evaluate the potential of milk runs assuming that the order policy can be adapted to delivery profiles. If the potential is high, the data collection process can be started. In this case it might be sensible to also consider inventory information from the suppliers because regular, levelled shipments allow for inventory reduction on the supply side as well (see Chapter 7).

10.2.3 Derived Input Parameters for the MRD

From the transport data analysis, four input parameters for the MRD can be derived for every supplier relation: a frequency per week, a shipping weight (in kg), a set of weekly patterns, and the AF cost for the corresponding weight and distance class. These data form the order of the MRD problem described in Chapter 9.

Frequency Assuming that the average frequency is economically suitable for the current AF tariffs, we can derive a minimum frequency from the historical transport data: We round the average frequencies per week for the daily and weekly suppliers to their next feasible frequency. Since there are only two supplier relations with more than two shipments per day, we restricted the study to a set of feasible frequencies of $\{1, 2, 3, 4, 5\}$.

Weight The weight per shipment is calculated by dividing the average weekly demand by the assigned frequency.

Weekly Pattern Weekly patterns or shipment profiles are considered as feasible if they fulfil the optimality property of their corresponding inventory model type from Chapter 7.

AF Cost Since the complete tariff tables, which were valid during the record period, are not available for the case study, the cost for an AF shipment with the assigned shipment weight needs to be estimated for every supplier plant relation. This was done in two different ways: In the first case we used a very simple approximation procedure, while in the second case we applied a commercial tool from the LOCOM GmbH for estimating tariff rates.

In the simple procedure we iterated over all shipments on the corresponding supplier plant relation and assigned the AF cost of the shipment which had the smallest absolute weight difference to the assigned weight. This procedure yields good results if the weights are not far away from the original values.

Since in one of the following scenarios the shipment sizes differ significantly from the original ones, we used a commercial tool of the LOCOM GmbH to estimate a full tariff table. The inputs for this tool are historic transport weights and costs and the distance of the supplier to the destination. The result is a tariff table or matrix with weight classes as one and distance classes as another dimension. The matrix is estimated in a way that the difference to the historic cost are minimal on the one hand and that a degressive cost structure for both dimensions is guaranteed[4].

10.3 Scenarios and Models

Based on the results described in the preceding sections we developed different test scenarios. In order to get a broader test set, we did not restrict our tests to this use case but generated additional scenarios relying on the real-world structure of this data base by varying some parameters.

The base case for all scenarios corresponds to the current situation at Bosch Automotive: All shipments are operated through the AF network. The cost for this base case is used as base line. The other extreme in terms of the transport concept allocations is the exclusive milk run case. In this variant all shipments for daily and weekly relations are executed by milk runs. An exception are the two suppliers of the frequency group daily and weekly (see Figure 10.4), which are located more than nine truck hours away from the plant in Homburg. We assume that these suppliers are operated through the area forwarding network in all cases.

10.3.1 Parameters

The different considered scenarios arise from the parameter variations described in this section:

CV Classes We vary the number of supplier relations considered in the MRD problem and, hence, the number of suppliers being candidates for milk runs. We distinguish four scenarios: The first class includes only the regular and well levelled suppliers with weight and frequency CVs smaller than 0.3. In the second class we include all relations with CVs smaller than 0.75 and in the third class all with CVs smaller than 0.9. For the last class the threshold is 1.33. Since none of the relations of

[4] For detailed information see http://www.xcargo.de/en/at-a-glance-transport-rate-calculation/ (14.02.2015)

our use case exceeded the latter threshold, the last CV class includes all daily and weekly relations.[5]

Table 10.3 shows the number of relations and the number of shipments included in the four different CV classes. The biggest instance with all daily and weekly suppliers contains 54 relations. Remember from Figure 10.3 that these suppliers represent only 22% of all suppliers but that 80% of all shipments, 89% of the total weight and 85% of all cost arise on these relations.

Table 10.3: MRD instance characteristics by CV classes

CV class	≤ 0.3	≤ 0.75	≤ 0.9	≤ 1.33
number of relations	8	39	44	54
number of shipments	38	129	140	154
share of aggregated weight	18%	42%	73%	89%
share of total AF cost	20%	55%	74%	85%

Vehicle Classes The analysis of the section before showed that the average shipping weight is rather low in this use case: Up to 30 suppliers would need to be served on a tour of a 40-ton truck to use up its weight capacity limit. Therefore, we investigate if using smaller and less expensive vehicles is beneficial: Table 10.4 shows the necessary parameters for three typical vehicle classes. Please be aware that the vehicle class BIG corresponds to the 40-ton truck introduced in Section 10.1.

Weight Classes In order to analyse how the models behave in cases in which the average weight of all shipments is higher, we add an artificial scenario: We simply multiply the shipment weights for all orders by a factor of 2 and use the estimated tariff table to assign corresponding AF cost to every order.

[5] As mentioned above with the limits we follow the classification scheme of (Hopp and Spearman 2001a) determining low ($cv \leq 0.75$), moderate ($0.75 \leq cv \leq 1.33$), and high variability ($cv \leq 1.33$). In order to get a broader set of instances we added the class ≤ 0.9.

Table 10.4: Vehicle parameters by vehicle class

Vehicle Class	Capacity [ton]	Cost per km [EUR]	Cost per hour [EUR]	Mark up
SMALL	1.95	0.3296	23.3955	0.155
MIDDLE	12.00	0.4100	31.3007	0.155
BIG	20.00	0.6100	34.7380	0.155

To assure that the two AF cost estimation strategies do not lead to significantly differing results, we test them both. Therefore, we consider three different weight/AF parameters: ($w = 0$) assigned weight with simple AF cost estimation, ($w = 1$) assigned weight with AF cost from the estimated tariff table and ($w = 2$) twice the assigned weight with AF cost from the estimated tariff table.

Table 10.5: Volume per week and AF cost of the (pure AF) base case by CV class and weight class

		CV class			
	weight/AF class	≤ 0.3	≤ 0.75	≤ 0.9	≤ 1.33
vol [ton]	w=0	25	60	106	129
vol [ton]	w=1	25	60	106	129
vol [ton]	w=2	51	120	212	258
base case AF cost	w=0	1	1	1	1
base case AF cost	w=1	1.03	0.93	1.07	1.06
base case AF cost	w=2	1.65	1.58	1.70	1.71

Table 10.5 shows the total volume per week considered in the following planning steps differentiated by CV and weight classes. Since the weight of all suppliers is simply doubled, the total weight in class $w = 2$ is the double of the total weight in class $w = 0$ and $w = 1$. The objective function of the base case can be calculated by summing up the AF cost for

all shipments. To keep sensitive cost values confidential, we give this base case cost as 1. In the following all cost are given relative to this base value.

The objective value differences between the $w = 0$ and $w = 1$ scenarios are between 3% and 7% deviating in both directions. This gives a first hint that the estimated tariff table does not systematically over- or underestimate the AF cost compared to the simple AF cost assignment heuristic. The cost increase factors of around 1.6 to 1.7 for double the weight in the $w = 2$ scenarios show the degressive cost structure of the tariff table.

10.3.2 Milk Run Design Models

The key goal of this case study is to identify the potential of milk runs as an alternative to AF shipments keeping frequencies from historical transport data roughly stable. Hence, we solve the TC-MRS problem introduced in Chapter 9. Since transport data usually do not contain any information about the parts being shipped, it is not possible to include inventory effects in this study. However, inventory cost compared to transport cost are usually small or even neglectable (compare for example case studies of (Chuah and Yingling 2005) or (Ohlmann et al. 2008)), therefore this should not reduce the value of the results significantly.

We assume that delivery profiles along with a levelled use of transport lots are wanted since this setup offers the most benefits if transport processes and cost are the main focus (see Chapter 2 and Chapter 8). Therefore, every feasible weekly pattern must fulfill the Property 7.6.1 of the CDI-LT variant.

Since today the arrival or inter arrival times are not restricted, we do not consider scheduling aspects on the level of hours in this case study. We only restrict the shipment patterns to the inventory optimal ones. The experiments of Chapter 8 showed that considering ATC increases the transport cost only marginally but leads to prohibitively long run times. For the bigger CV classes it would make the use of a heuristic unavoidable. Keeping in mind the very small cost increase for ATC compared to NOTC variants, we leave out the scheduling aspect favouring optimality for assessing the model effects without needing to consider a bias of heuristic solutions.

We assume that pick up times of milk runs resulting from the milk run planning step are accepted by the suppliers – of course, respecting their opening hours. This is not a very strong assumption since it is rather an improvement compared to the situation of today, in which the pick up time is arranged only the day before the delivery between the logistics provider and the supplier.

The data analysis showed that vehicles with smaller capacities might be beneficial for a milk run system. Hence, we consider a heterogeneous fleet with differing capacities and differing cost parameters not restricting the number of vehicles per type.

From the situation today it can be deduced that Bosch does not want to operate the milk runs using its own fleet but that they outsource them to a logistics provider. Since on the way back to the depot the logistics provider can acquire return freight, a milk run tour in the case study does not include the way back. Furthermore, since we do not have any vehicle depot information of the subcontractors, we assume that a tour starts at the first supplier and ends at the Bosch production site, that means we solve an open VRP (see Section 5.3.1).

Formally spoken, we apply the TC-MRS model from Chapter 9 without consistency extensions but considering different vehicle classes and open tours. We, furthermore, compare the TC-MRS solutions to results from applying the pure milk run model (MRS) of Chapter 8 and the weight based allocation heuristic (WA) introduced in Chapter 9.

10.4 Computational Experiments

For implementing and solving the set cover model we used Gurobi 5.6.3 with default settings accessed via its Python 2.7 API. Since we do not consider scheduling aspects, we have no temporal synchronisation between tours and, consequently, only need to add open tours optimal with regard to distance and driving time to the set of milk run candidates. To generate them, we implemented a dynamic programming based algorithm in Python 2.7.

We solved 360 instances – resulting from the parameter variations and model variants – on two common office machines. All TC-MRS variants were solved on a machine with Intel Core i7-260M, 2.8 GHz, 8.0 GB RAM with a 64 bit Windows 7 operating system. The MRS and WA models were solved on a PC with an Intel Core i5 760, 2.79 GHz, 8.0 GB RAM and with a 64 bit Windows 7 operating system. We were able to solve all instances to optimality: That means we generated all feasible, open tours ending at the Bosch plant and yielding the shortest distance for the given set of suppliers, and we solved the set cover to optimality.

We split the analysis in three phases: The first phase is dedicated to identifying the potential of milk runs as an alternative to the AF system, the second phase focuses on the effect of additionally considering a heterogeneous fleet, while the last phase is dedicated to show the advantage of a TC-MRS based transport concept decision compared to a WA heuristic as used in practice.

10.4.1 Phase I: TC-MRS

In phase I we focus on the impact of different CV classes, vehicle and weight types on the savings potential of milk runs. Table 10.6 summarises the different parameter variations considered in this phase.

Table 10.6: Parameter variations

CV class	Vehicle class	Weight class
≤ 0.3	SMALL	w=0
≤ 0.75	MIDDLE	w=1
≤ 0.9	BIG	w=2
≤ 1.33		

In the following we give the parameter set of a scenario as a triple in the order (CV class, vehicle class, weight class).

MRS In the first step we compare the exclusive milk run variant (MRS) to the base case: Figure 10.7 shows the cost compared to the AF base for different scenarios. Please be aware that in this chart only the cost for the relations which are part of the instance are considered.

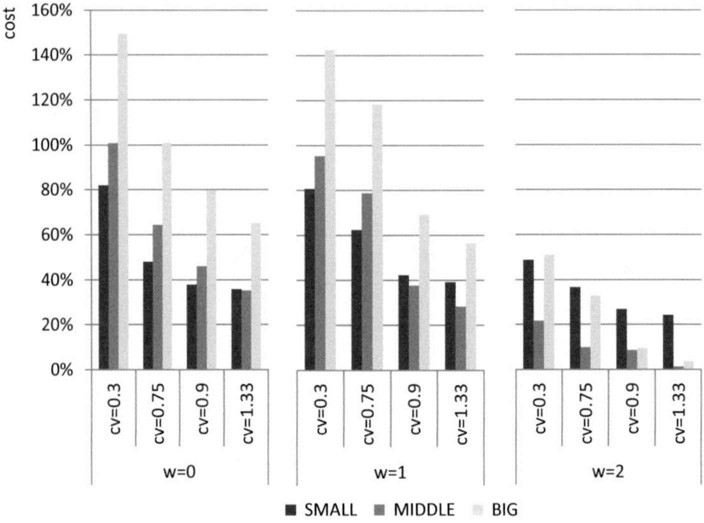

Figure 10.7: Cost compared to the base case for the MRS model variant

Especially for the smaller instances cv=0.3 and cv=0.75 with 8 and 40 suppliers respectively, a pure milk run system is a very expensive alternative. For the typical 40-ton vehicles (BIG) the cost increase is partly above 140%. The shipment structure with very light loads lead to milk runs with a bad capacity usage rate. This becomes significantly better the bigger the instances and, hence, the bigger the potential to build cost efficient tours are. However, for the two original weight variants ($w = 0$ and $w = 1$) the usage of the AF network with the current tariff tables lead for all scenarios to significantly lower transportation cost than an exclusive milk run system.

The scenarios of $w = 2$ have a much better weight structure for milk runs, at least if the whole optimization potential can be used for building efficient tours: The cost increase with the vehicle type MIDDLE is for the two bigger instances with 45 (cv=0.9) and 52 (cv=1.33) suppliers only 9% and 1% respectively. For the latter scenario, this means that a milk run system planned by the receiving plant would yield cost only one percent above the cost of the AF system. This result suggests that the AF tariff table for the weight structure $w = 2$ is significantly worse compared to the $w = 0$ and $w = 1$ scenarios.

Even in the pure MRS case, there appear small cost differences between the scenarios of $w = 0$ and $w = 1$ in which only the AF costs differ: This can be explained by the group of suppliers which are serviced through the AF network since their load exceeds the capacity of the vehicle types SMALL and MIDLLE or since they are further away from the plant than 9 hours truck driving time. However, since the results are similar we focus in Figure 10.8 on scenario $w = 1$ and $w = 2$.

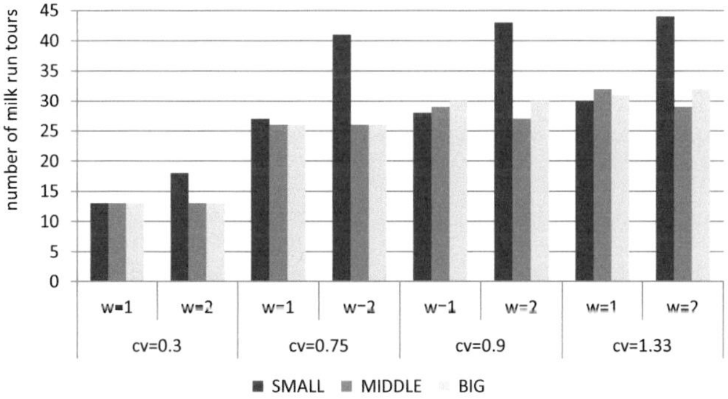

Figure 10.8: Number of tours for the milk run only scenario

This chart reconfirms that the weight structure is much better in case w=2: Even if double the weight is transported in $w = 2$, the number of tours compared to the $w = 1$ scenario increases only marginally for the vehicle classes MIDDLE and BIG and moderately if SMALL vehicles are considered. That means the vehicle capacity usage increases considerably as shown in Figure 10.9. Simultaneously the time capacity usage stays roughly constant over all scenarios apart from $w = 2$ with SMALL vehicles. This means that apart from the latter in all scenarios the time capacity is binding. Or stated differently, in our case study the number of stops per tour is in most cases restricted by the maximum working time of the driver and not by the weight capacity of the trucks.

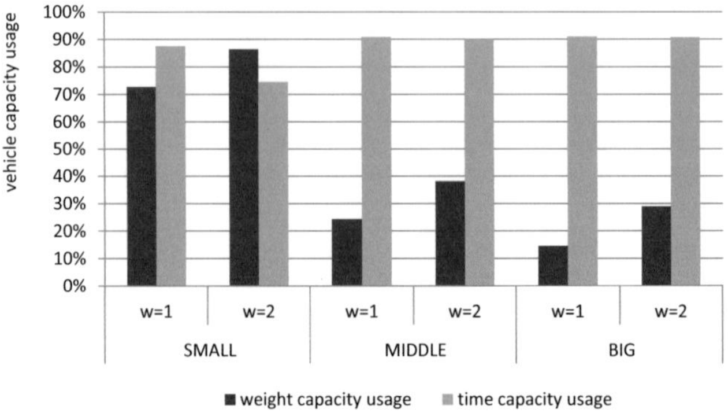

Figure 10.9: Average time and weight capacity usage over all CV classes for the milk run only scenario

It might be surprising that not in all cases the number of tours decreases with an increase in vehicle size (see for example scenario w=1 and cv=0.9): This can be explained by the fact that there exist orders which cannot be operated by the two smaller vehicle types because their weight is bigger than the capacity of the vehicle. Since these relations are operated by the AF network, fewer stops are served by milk runs. For the scenario (w=1, SMALL, cv=0.9) for example 39 orders are operated by milk runs, while in the same scenario with vehicle type MIDDLE or BIG 43 orders are served by milk runs.

Another not intuitive effect is that in the scenario (cv=0.9, MIDDLE) and (cv=1.33, MIDDLE) the number of tours for $w = 1$ is smaller than for $w = 2$. This can be explained by the effect that there are two orders in the $w = 2$ scenario whose demands exceed the vehicle capacity of 12 tons. Hence, these orders are assigned to the AF network and less tours are needed in the w=2 scenario.

TC-MRS A first glance at Figure 10.10 shows that the resulting objective values of the TC-MRS model – the sums of AF and MR costs – are very similar for the two AF cost estimation methods labelled by w=0 and w=1. Since both estimation methods lead to similar results, we focus in the following on the two more comparable weight classes w=1 and w=2.

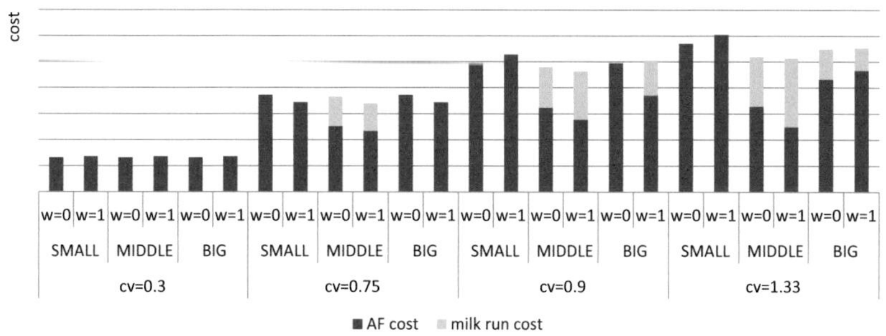

Figure 10.10: Absolute AF and milk run cost in Euro by scenario (for reasons of confidentiality the y-axis labels are left out)

Figure 10.10 furthermore shows that there is no scenario in which orders are allocated to the milk run transport concept if only SMALL vehicles are available. This is also true for the w=2 scenarios. Hence, in the following we restrict the analysis on the scenarios with the vehicle classes MIDDLE and BIG.

In Figure 10.11, especially the results of the scenarios w=1 reconfirm the results of the section before: The bigger the instance, the higher is the potential for cost efficient milk runs and, hence, the potential for savings by applying the milk run concept. In scenario cv=1.33 the potential savings are up to 15% and 18% for the w=1 and w=2 case respectively. Keeping in mind that the transport cost for serving the 54 daily and weekly supplier relations – corresponding to the cv=1.33 scenarios – represents more than 85% of the total cost, this decrease in cost is significant.

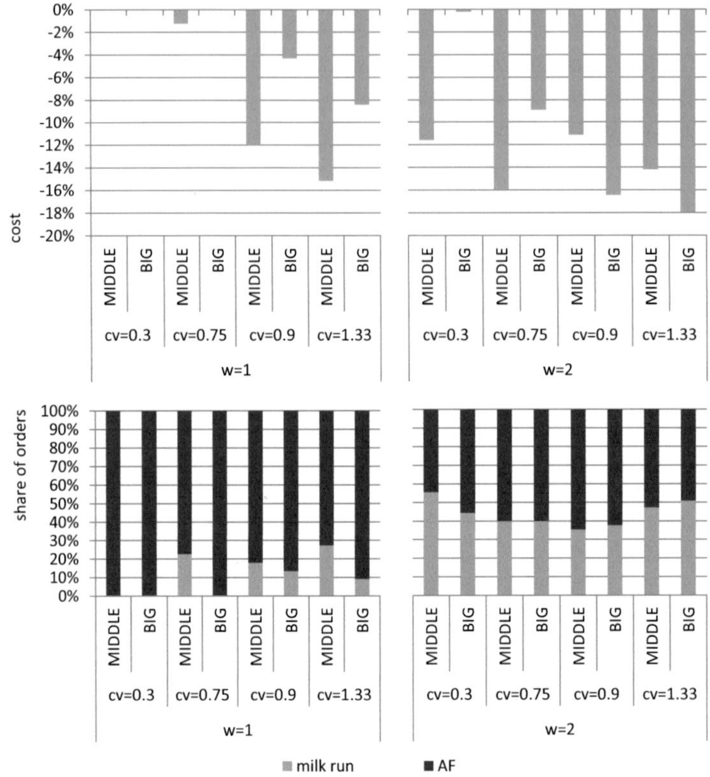

Figure 10.11: Cost compared to the base case for the TC-MRS model results (upper charts) and proportion of milk run orders and AF orders (lower charts)

Considering change over cost, it is interesting that the cost saving of 12% and 15% for (cv=0.9, MIDDLE, w=1) and (cv=1.33, MIDDLE, w=1) respectively, are reached by shifting only 18% and 27% of the orders to the milk run transport concept. That means that only 8 and 15 relations respectively need to be reallocated to the milk run concept to reach this high cost savings.

If in the $w = 1$ scenario the supplier relations considered as potential milk run candidates are restricted to the ones with a low variability (≤ 0.75), the savings are neglectable. Please remember that the saving of 1% is relative to the cost of the instance, which represents only 55% of the total cost (see Table 10.3).

In case of the $w = 2$ scenarios the savings also for the small instances are significant. Even if just the well levelled relations are considered as milk run candidates, the transport concept is beneficial. Furthermore, in all scenarios – even in the (cv=0.3, BIG, w=2) with a savings potential below 1% – the proportion of milk run orders is much higher. This is also reflected in the charts of Figure 10.12 showing the number of milk run tours per week: The number of milk runs, which would be operated in accordance to the TC-MRS results, lies between 5 and 18 over all scenarios.

It is remarkable that the selection of the vehicle type not only depends on the weight structure but also on the number of relations considered: For the scenarios (w=2, cv=0.3) and (w=2, cv=0.75) the vehicle type MIDDLE yields lower cost, while for the two bigger instances (w=2, cv=0.9 and cv=1.33) the BIG vehicle type is more cost efficient.

Comparing the capacity usage of the MRS scenario in Figure 10.9 and the capacity usage of the TC-MRS solutions in Figure 10.13, the effect described in Chapter 9 is reconfirmed: Especially the capacity usage measure of the non restricting dimension – the weight capacity in our case – improves if the area forwarding network is available as an alternative for the suppliers not fitting to a cost efficient milk run in terms of weight or geographic position. On average, the capacity usage for MIDDLE and BIG vehicles increases from around 30% in the milk run only scenario to almost 56% in the TC-MRS model results. The time capacity usage is stable – above 80% on average – compared to the milk run only results.

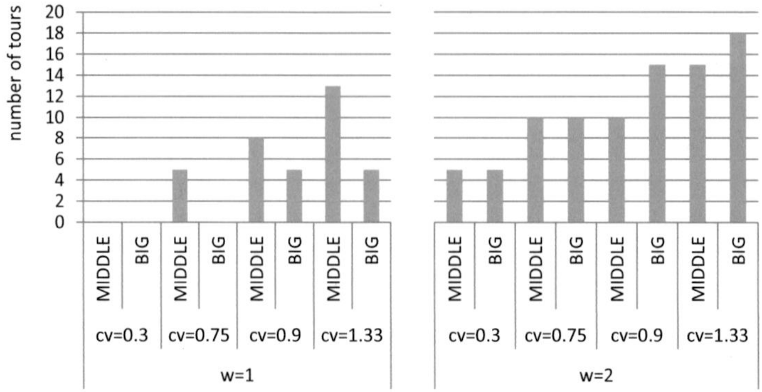

Figure 10.12: Number of milk run tours

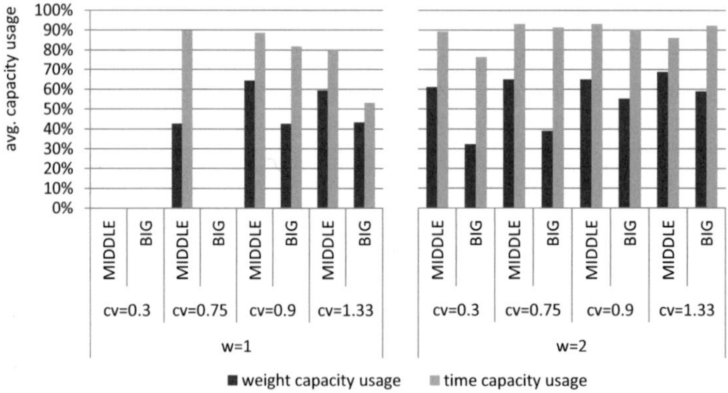

Figure 10.13: Average time and average weight capacity usage over all resulting milk run tours

10.4.2 Phase II: TC-MRS with a Heterogeneous Fleet

The results of phase I show that the selection of the right vehicle type is difficult and does not only depend on the weight structure. Hence, in phase II we consider a heterogeneous fleet of vehicles and leave it to the model to select the milk runs with the corresponding vehicle type. The vehicle class yielding minimum cost for a tour can be determined in advance since we do not restrict the number of vehicles available for each type. Hence, the increase in size of the set cover model is neglectable.

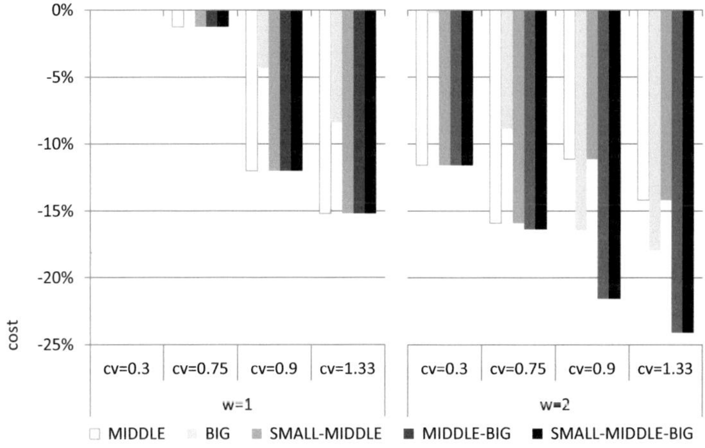

Figure 10.14: Cost compared to the base case for the TC-MRS model results for different sets of vehicle types

Figure 10.14 shows that for the scenarios of weight class 1 the savings cannot be increased by considering different vehicles. The highest saving value can be realised by planning with vehicle type MIDDLE.

However, the scenarios of w=2 show a significant potential increase in savings for the two bigger instances cv=0.9 and cv=1.33: The increase from giving a set of vehicles of size MIDDLE and BIG into the decision model instead of just the best type is around 5 percentage points (to 21%) and 6 percentage points (to 24%) respectively. Having additionally the vehicle SMALL available does not improve the situation under the given circumstances.

The additional savings result from different effects: Either similar tours of the BIG vehicle are taken over by the smaller and cheaper vehicles of type MIDDLE – which results in a better capacity usage measure. Or more orders can be assigned to the milk run concept since more tours are cost efficient. Both effects occur in the cv=0.9 and cv=1.33 case (see Figure 10.15).

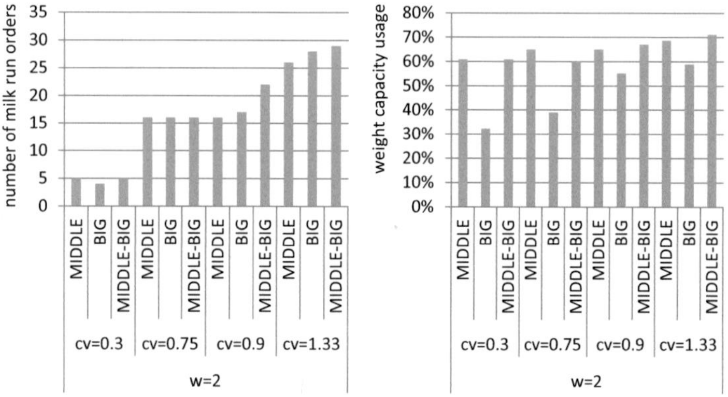

Figure 10.15: Number of milk run orders (left chart) and average weight capacity usage over all milk run tours by scenario

In the case cv=0.75, the vehicle type MIDDLE is the most cost efficient one and leads to the biggest savings. The scenario with MIDDLE and BIG vehicle types available leads to a slightly greater saving but to a worse capacity usage. This is possible since the bigger vehicle took over just one tour with one stop interchanged, which is shorter and hence cheaper even with the more expensive BIG vehicle.

The results of this phase showed that it is worth to consider different vehicle categories when solving the TC-MRS. This is reconfirmed with a look at the MRS variant combined with a heterogeneous fleet for the w=2 scenario: Figure 10.16 shows that the cost increases for a pure milk run system are considerably smaller for the small instances and even turn into significant cost savings compared to an exclusive AF system for the bigger instances. In the latter cases, it is beneficial to have the whole set of vehicle types – also the SMALL ones – within the set of available vehicles.

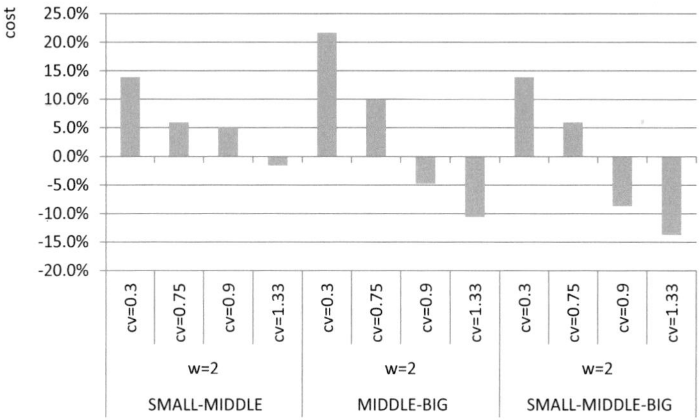

Figure 10.16: Cost compared to the base case for the milk run only model variant considering different sets of vehicle types

10.4.3 Phase III: TC-MRS vs. WA Heuristic

In phase III we analyse the advantage of the TC-MRS compared to the weight allocation heuristic introduced in Chapter 9. As described before, the WA heuristic allocates supplier relations to a transport concept depending on their order weight. That means, all orders below a threshold wa are assigned to the AF concept, and all above are served by milk runs.

For the analysis we tested three different values for wa, namely 250 kg, 500 kg and 1000 kg. In both model variants – WA and TC-MRS – we consider the set of all vehicle types in order to compare the smallest possible cost in both cases.

The charts of Figure 10.17 confirm the results of Chapter 9: Depending on the size and the weight structure, a significant cost increase of up to 60% compared to the base case occurs if the WA heuristic is applied. At the same time, the results of the heuristic are strongly deviating both over instance size and weight classes.

In 5 out of these 8 scenarios it is beneficial to apply the WA heuristic since it is possible to reach significant cost savings of up to 11% in the w=1 scenarios and 18% in the w=2 scenarios. However, a proper choice of the threshold is crucial for yielding these results, with none of it dominating the others. Hence, different thresholds need to be tested for realising these savings. But a manual planning process for the MRS is prohibitively time consuming. Furthermore, it is worth to leave the transport concept decision as optimisation potential to the model since for the two big scenarios another 4 percentage points – summing up to 15% – and 6 percentage points – summing up to 24% – savings respectively can be realised. The biggest difference between the best WA model result and the TC-MRS result even reaches 12% in case of the (cv=0.9, SMALL-MIDDLE-BIG, w=2) scenario.

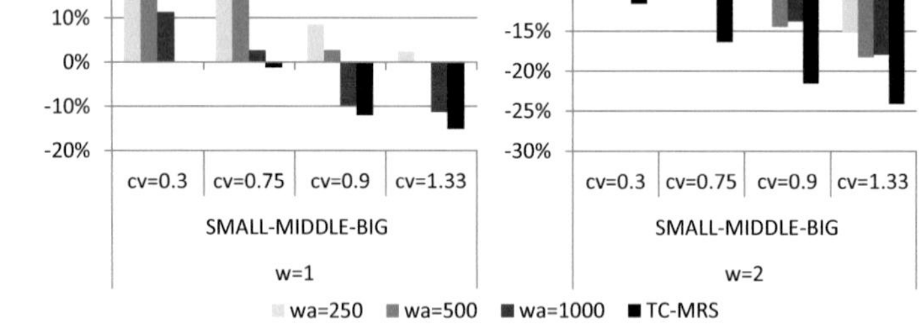

Figure 10.17: Cost compared to the base case for WA model variants and the TC-MRS considering the set of all vehicle types

10.4.4 Run Times and Limits of the Approach

As already mentioned, the TC-MRS – using the a priori column generation approach – could be solved to optimality for all scenarios of our use case. However, the approach has its obvious limitations with respect to the number of orders. The limits are influenced by the maximum number of stops per tour, which in turn is restricted by the weight and working time limitations and, hence, by the weight structure and the geographic distribution. In the following we analyse the impact of the different parameters on the run times of the two stages of the approach – the a priori column generation phase and the set cover phase.

Generating Columns

Figure 10.18 shows the run times for generating the columns over the number of orders, that means, for the four different instance sizes defined by the cv thresholds. As mentioned, in this step we enumerate all feasible and shortest tours. Obviously, the very restrictive weight limit of the SMALL vehicle causes shorter run times since the search can be pruned heavily by the capacity limit. Since the weights in scenario w=2 are bigger, the run times are even shorter. In case of the midsized and big vehicle types the pruning mainly result from the maximum tour duration and hence the run times are almost the same.

The longest resulting tours have 3 to 4 stops in the smallest scenario and between 8 and 9 stops for all bigger scenarios. The total numbers of resulting tours is given in Figure 10.19. Analogue to the run times, they strongly vary between 40 for cv=0.3 and up to 200,000 tours for the cv=1.33 scenarios.

In general, the column generation run times for the biggest instances of our use case are around 10 minutes if the number of stops per tour is not restricted at all. With a restricted number of stops per tour the run time decreases significantly. For the biggest instance with the following parameter set (vc=1.33, BIG, w=1) it drops from 600 to 300 seconds if the number of stops is restricted to 6, and to 70 seconds if it is restricted to 5. This is interesting since in practice more than 5 or 6 stops on a milk run tour are often not accepted.

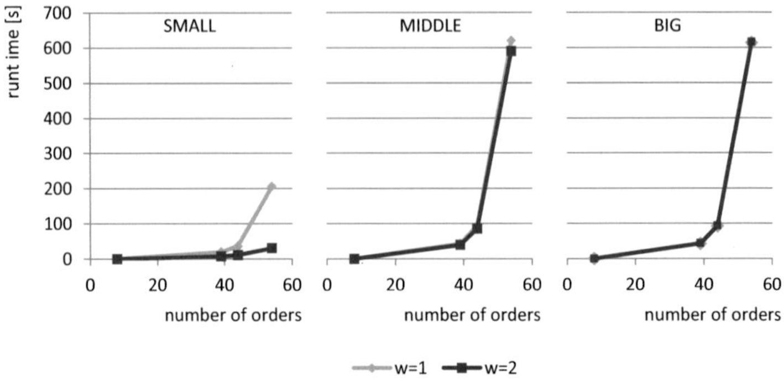

Figure 10.18: Run time of the column generating phase with the number of orders on the x-axis differentiated by the three vehicle types SMALL (left), MIDDLE (middle) and BIG (right)

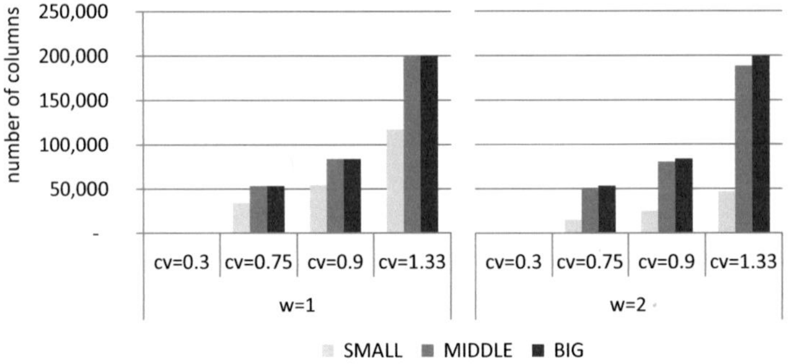

Figure 10.19: Number of columns by scenario

Solving the TC-MRS Set Cover Model

For the set cover phase the number of columns plays a decisive role in terms of solution time: In Figure 10.20 the average run times over all vehicle sets are illustrated differentiated by instance size and weight class. All scenarios were solved to optimality with an allowed relative MIP optimality gap of 0.0001 – which corresponds to the default value of the applied Gurobi version.

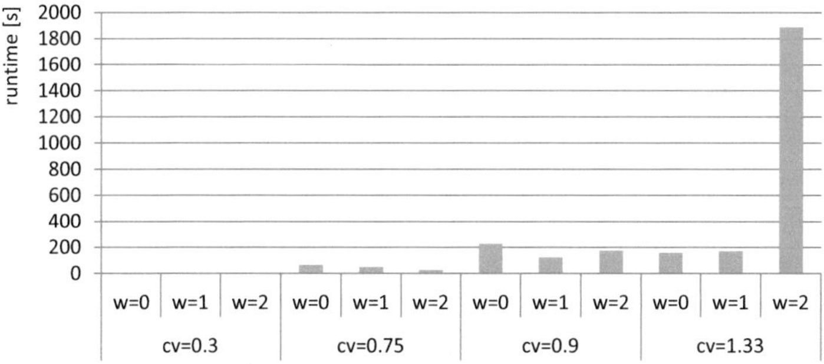

Figure 10.20: Average run time of the set cover solution phase over all vehicle types

Another factor which seems to influence the run times of the set cover phase is the resulting share of milk run orders. The more orders are shipped through the AF network the smaller is the probability for long run times: In case of the cv=1.33 scenarios, the average run time for the w=1 scenario is around 170 seconds and the average proportion of milk run orders is 20%. The average run time for the w=2 case is around 1900 seconds – even if in the w=2 case 1% fewer feasible tours are considered. However, the resulting average share of milk run orders is above 40%. All scenarios which result in an AF share of 100% are solved within less than a minute of run time. This tendency is confirmed by the fact that the milk run only scenarios cause the longest run times.

Limits of the Approach

Even if there is some room for improving the performance of our algorithm for the a priori column generation phase, the general picture is clear: The run time grows exponentially with the number of customers (see Figure 10.18). As shown, the maximal number of stops – either restricted by the vehicle capacity or restricted due to operational reasons – influences the run time significantly and can be used to shorten run times. Since the number of tours also impacts the run time of the second stage, the set cover phase, the run time decrease is doubled.

In cases in which the number of regular suppliers is a few hundreds, a heuristic approach needs to be applied. On the other hand – besides the case study on hand – the literature shows that there exist other use cases where the proposed approach, yielding optimal results, seems to be applicable: Alegre et al. (2007) solve a long-term PVRP for an auto part manufacturer incorporating 72 regular orders from 12 locations. Peterson et al. (2010) report on a planning problem of weekly milk runs for a distribution network in North America of a home appliances producer incorporating 75 regular clients. The daily milk runs of a Toyota plant in (Chuah and Yingling 2005) contain 24 locations, while Ohlmann et al. (2008) report on real-world data of Toyota containing between 13 to 109 daily supplier relations.

10.5 p-ratio Indicator

Especially for small and medium sized companies it might be beneficial to have an easy to calculate parameter indicating if weekly recurring milk runs are a cost efficient alternative to an area forwarding system in place. This would be even more interesting if one could "play around" with the expected vehicle cost for milk runs.

In order to judge if a supplier relation is a potential milk run candidate, one needs to trade off the AF against the milk run cost. The AF cost c_i^{AF} are easy to determine since they only depend on the distance and weight class the order falls into. However, the milk run cost c^{MR} depends on the total distance and operation time of the tour and can be only determined if the tour is completely known. Hence, for the latter we determine an upper bound $\bar{c}_i^{MR}$ by assuming a supplier is served on an exclusive milk run tour. These costs can be easily calculated by determining the driving distance and operation time and multiply the values with the respective vehicle cost parameters. If the weight capacity or the maximum tour duration is exceeded, the cost is set to infinity.

Using these cost parameters we can determine the p-value – already introduced in Chapter 9 for determining consistent AF cost parameters – for each supplier relation i and for each vehicle type v as follows:

$$p_i^v = \frac{c_i^{AF}}{\bar{c}_{iv}^{MR}} \tag{10.1}$$

We further approximate the milk run cost by assuming that the cheapest feasible vehicle is chosen for the exclusive milk run and, hence, state:

$$p_i = \frac{c_i^{AF}}{\min\limits_v(\bar{c}_{iv}^{MR})} \tag{10.2}$$

The interpretation of this p-ratio is easy: Whenever p is bigger than one, the exclusive milk run is $(p_i - 1) \cdot 100\%$ cheaper than the AF shipment and, hence, the introduction of a milk run generates savings. Even more savings can be generated if on this milk run other suppliers "on the way" to the plant are serviced.

If a supplier has a high p-ratio just below 1, the probability that a cost efficient tour including this supplier along with other stops on the way can be built is high. Other suppliers combined to a tour do not need to necessarily have high values of p, they must rather be geographically well located. This becomes clear with a look at the charts of Figure 10.21: We show the values of p on the y-axis and all suppliers on the x-axis, sorted by p and differentiated by the assigned transport concept in the optimal solution for two scenarios. The orders with the highest p-ratios are part of the milk run tours. They are combined with some suppliers with a relatively low value. This is representative for all scenarios for which a cost saving could be achieved by applying the milk run concept.

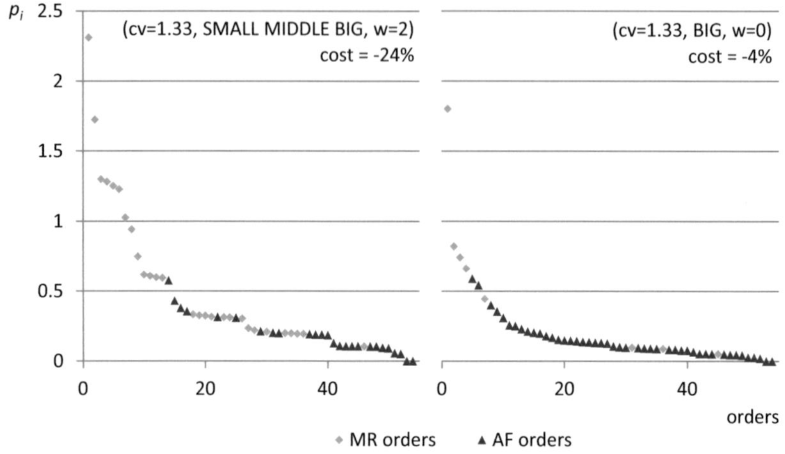

Figure 10.21: p-ratio for all suppliers sorted by their p-ratio of two representative scenarios with potential cost savings compared to the base case

The upper chart of Figure 10.22 shows the relationship between the average p-ratio over all supplier relations and the resulting TC-MRS cost differentiated by the three weight classes. As expected, high savings can only be realised for instances with a high average p-ratio. The other way around high average ratios do not guarantee savings.

In the lower chart of Figure 10.22, we put the maximum p-ratio instead of the mean value on the x-axis. This chart confirms that there exists a clear relationship: There is no case amongst the 72 scenarios in which savings can be realised if the maximum p value is below 0.8. This shows that the maximum p-ratio is a better indicator for the potential of milk runs than the average p-ratio.

From the tests in this case study we cannot deduce in general that milk runs might not be beneficial in scenarios with a lower maximum p-ratio. For other geographical situations than the one considered here, this threshold might be lower. It is an interesting question if a range for this threshold can be deduced by investigating a wide range of different use cases with other geographical characteristics. However, the case study shows that high maximum p-ratios close to one or even above can be considered as a clear indicator that the introduction of milk runs as an alternative to the AF network should be taken into account.

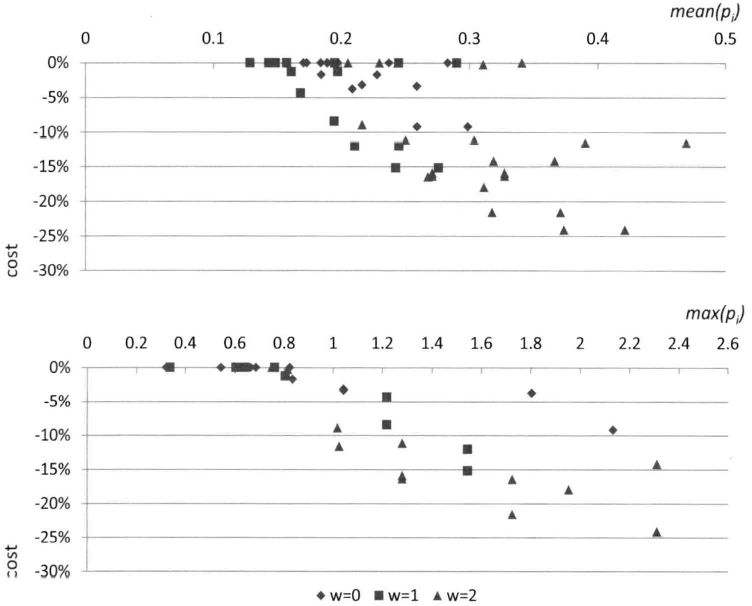

Figure 10.22: Cost compared to the base case depending on the mean (upper chart) and maximum p-ratio (lower chart)

10.6 Summary and Discussion of the Results

After summarizing the most important insights of the case study we conclude this chapter by critically discussing the results with respect to their impact on the planning situation of the receiving plant and the changes of the contractual situation with the logistics providers.

10.6.1 Summary of the Results

The analysis of the historical transport data showed, as expected, that weekly shipments are the central cost drivers within the inbound network of Bosch Homburg. Hence, weekly milk runs have the potential for cost savings and process improvements and, at the same time, reduce variabilty. However, one needs to keep in mind that there is an effort for planning and operating the milk runs, as well as for adapting the order policies to reach levelled frequencies and reasonably levelled shipment sizes. This needs to be especially emphasized since substantial cost savings of up to 15% for the original weight scenario can be only realised if most of the daily and weekly supplier relations are considered as milk run candidates.

Even if this could not be shown in the problem instances of the original weight classes $w = 0$ and $w = 1$, another important insight is that it is worth to solve the TC-MRS considering a heterogeneous fleet: In the weight class $w = 2$ – which has in general more beneficial weight structures for milk runs than the other two – the additional savings added up to 12 percent points compared to the best cases considering just one vehicle type. And this effect is even stronger in a pure milk run system: If a heterogeneous fleet is considered, the pure milk run system caused less cost than the AF network base case for the biggest instances of all three weight classes.

Even if we produced optimal MRS results after allocating the transport concept in accordance with the weight allocation heuristic – which is difficult to achieve if the tour planning is done manually – we showed that the heuristic is very sensitive to the wa threshold and does not reach the savings potential of the TC-MRS approaches. Hence, a weight based decision is not appropriate to exploit the potential of a milk run system.

The p-ratio indicator implicitly considers, besides the weight, the AF price depending on distance and weight classes, as well as an upper bound of cost for different vehicle types. It proved to be a reliable parameter indicating if an order on its own is a good seed stop for a milk run tour. Aggregated by a maximum function over all supplier relations it seems to be a good quick check to assess the attractiveness of the milk run system if an AF inbound system is in place. It is even possible to consider different realistic vehicle cost parameters.

If a milk run system is not considered as an alternative for operational reasons, the p-ratio can also be used to assess the tariff table for the typical shipment sizes of all suppliers. Instead of using estimated cost for milk runs, one uses estimated cost for dedicated ad hoc tours: If there exists a high share of relations with values close to one or above, this is a signal that the tariff table leads to rather high transport cost compared to planning tours ad hoc.

Independently from the concrete results of the case study, we could show that typically available historical transport data along with limited additional data can be used to generate input data for the TC-MRS model. Hence, the effort for a first analysis assessing the potential of cost savings by introducing milk runs is small. However, it can be an important step for arguing that the effort for the production and order planning to reach more levelled transport processes result in short-term savings. Especially, since the medium- and long-term advantages for reducing the variability of the whole supply chain are apparent.

The introduced a priori column generation approach clearly has its limits in terms of the instance sizes which can be solved in acceptable run times. However, we showed that there exist real-world use cases to which it is applicable and for which it yields optimal results. The tour generation phase just has to be done once and the resulting tours can be persisted. Then the user is able to do what-if analysis – for example, playing around with different AF or vehicle cost parameters – solving only the set cover model.

To the best of our knowledge, in literature there does not exist work on solving periodic vehicle routing problems considering a heterogeneous fleet with the possibility of outsourcing stops. If heuristics need to be developed to solve bigger instances, the a priori approach might be used to generate starting solutions by restricting the number of allowed stops as done in (Peterson et al. 2010). Furthermore, the p-ratio might be an interesting indicator for local search based heuristics – for example, for choosing seed values for insertion based construction heuristics.

10.6.2 Impact on Planning Systems

The potential cost savings and other advantages, such as complete transparency for the transport processes, are opposed by the increased planning and operating effort for these milk runs for the receiving plant. For planning and operating the tours – even if they are operated by vehicles of a logistics provider – an IT-support is necessary: In the use case, the number of resulting milk runs went up to 20 per week in a mixed system and up to 44 in a pure milk run system. They need to be planned and re-planned if input parameters change. These tours need to be operationalised on a daily base and they need to be monitored in order to identify the moment when they are no longer cost efficient. In an AF inbound system, these steps are executed by the logistics provider and are a black box for the receiving plant.

For monitoring the cost efficiency of milk runs in a mixed system it is possible to use the p-ratio on the tour level for the actually shipped volumes: Assuming that there exists one milk run, one calculates for the actual orders the virtual AF costs for all stops and relates them to the actual milk run costs. If the ratio is above one, the milk run is cost efficient compared to the AF network.

If the vehicle capacity is not exceeded, the actual milk run cost corresponds to a fixed value. Otherwise, there are different possibilities to react: We assume for the following example that an order can be split and, hence, the exceeding volume can be shipped through the AF network. This means that the total cost for the shipments of all suppliers of a milk run tour are usually below the cost for the pure virtual AF shipments since the cost advantage of the milk run is still existing. An exception is the case in which the AF cost are strongly degressive with increasing volumes. The other way around, if the shipment volume is significantly below the average value, the AF cost can fall below the fixed cost of the milk run.

These relationships become clear with a look at the simplified, artificial example of Figure 10.23: We consider a milk run of three stops A, B, and C operated by a vehicle with a capacity of 60 volume units at a price of 78 cost units. The AF cost are assumed to be 1.4 cost units times the shipping volume. On the left column we show a scenario in which the

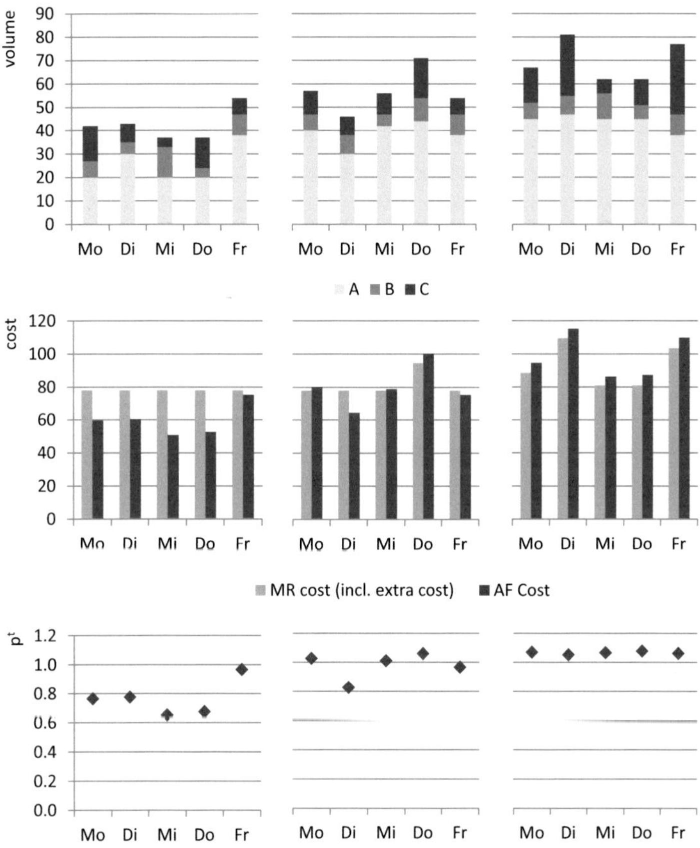

Figure 10.23: Volume (upper row), cost (middle) and p-ratio on tour level (lower row) for different demand realisations: low demand (left column), normal demand (middle) and high demand (right column)

demand realisation is smaller than expected, in the middle it corresponds to the expectation, and in the left column it is higher. Whenever the capacity is not exceeded, the milk run costs stay constant at 78 cost units. In the other cases, the extra AF cost for the exceeding volume is added to the actual milk run cost showed in the charts in the second row. The virtual AF cost vary in all cases with the volume. The last row of charts shows p on the tour level. In the scenario of the low demand realisations, the execution through the AF network would be better in terms of transport cost. In the scenario of the high demand realisation, it becomes clear that the pure AF costs are usually not reached in a setup in which order splits are allowed. However, in the medium-term, it might be better to adapt the milk run if the demand is assumed to be permanently high because all other advantages of milk runs get lost and the system gets more complex to operate. If order splits are not allowed, the situation is completely different.

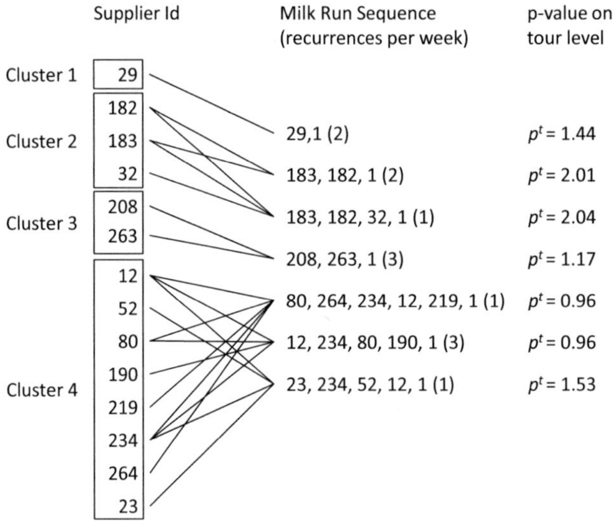

Figure 10.24: Milk run clusters and corresponding p-values on tour level resulting from the solution of the TC-MRS model for instance (cv=1.33, MIDDLE, w=1)

It is a matter of future work to investigate the impact of different types of recourse or repair strategies in case of capacity violations during the MRD. However, we can state that the p-value on tour level is a helpful indicator for measuring the cost efficiency of a running mixed AF and milk run system: It is better than measuring only the capacity usage of milk runs which is common in practice but which does not contain any information on the cost efficiency relative to alternative transport concepts.

Whenever suppliers are serviced in different milk runs with differing suppliers, it is possible that a single milk run has a p-value on tour level which is below one (see the tours of cluster 4 in Figure 10.24). That means, considering only such a milk run it would be better in terms of cost to assign the related suppliers to the AF network. However, considering the whole geographic cluster it is beneficial to assign all suppliers to the milk run concept. Hence, since there exist these cases, the p-value should not be calculated on tour level but for the whole cluster.

10.6.3 Impact on Contractual Situation

To conclude this chapter, we discuss the impact on the contractual situation between the receiving plant and the AF logistics provider: If the TC-MRS is solved, this can be considered as "cherry picking" from the perspective of a logistics provider. The favourable orders in terms of weight and geographic position are served by milk runs, planned and controlled by the receiving plant, while the unfavourable relations with most probably low volumes are left to the AF network. This means that the volume and, at the same time, the potential for building cost efficient tours in the affected area decreases for the logistics provider.

If the shifted volumes are relatively small compared to the total shipping volume of the receiving plant or compared to the total shipping volume of the logistics provider in the area, it does not affect the general cost structure of the logistics provider significantly and, hence, the AF tariffs offered to the receiving plant. However, if the volumes are comparatively high, the changed cost structure forces the logistics provider in the medium-term to raise the tariffs.

In such a case it might be of interest for both parties that the milk runs and the AF shipments are served by the same logistics provider: The volumes would not decrease and it would be even possible to leave it to the logistics provider to fill up a milk run tour with additional stops if it can be guaranteed that the pick-up and delivery time changes lie within a certain interval (compare to the master tour concept introduced in Section 2.1.2). In some situations this might combine the advantages of both systems: reliable milk run processes for the daily and weekly supplier relations and optimization potential for short-term or irregular shipments.

Another possible scenario could be that a logistics provider offers better conditions if the clients follow regular shipment profiles so that the logistics provider can offer milk runs with fixed schedules serving different clients. This provides greater optimization potential and reduces the cost and at the same time the variability in the processes.

However, it is beyond the scope of this work to determine how contracts must be designed to set the right incentives for the consignee and the freight forwarders.

11 Conclusions and Outlook

In this chapter we summarize the main contributions and outcomes of this thesis with respect to the goal from a practical point of view – to provide models and indicators to manage the planning complexity of milk runs – and with respect to the research questions formulated in Section 1.1. We conclude this thesis with an outlook on opportunities for future research.

11.1 Contributions and Results

From a practical point of view, the goal of this thesis was to provide models and indicators to manage the planning complexity of milk runs. Figure 11.1 shows the three main contributions of this thesis, and it shows which sets of research questions they address. Since there exists no unambiguous assignment of chapters to research questions and the contributions cannot be mapped to a distinct chapter or a single set of research questions, we use the three key contributions of Figure 11.1 as a structure to discuss and summarize the results of this thesis.

11.1.1 Definition of the Milk Run Design Problem

The first part of this thesis is dedicated to defining milk runs as a transport concept within the automotive industry. That means, we answer the research questions 1 and 2 from a conceptual point of view abstracting from the concrete use cases treated in literature.

Definition and Discussion of Road Transport Concepts

Due to the ambiguous use of the term milk run in practice and literature we proposed an extended classification scheme characterising different road transport concepts typically applied within the German automotive sector. In former work, only the physical network structure was considered as classification criteria. However, that is not enough to define the existing concepts. Therefore, we added the tariff structure – defined by the cost structure and the type of contract, the degree of consolidation, the planning horizon, and the responsibility for the tour planning task as further classification criteria.

Based on the discussion of the road transport concepts we defined the term milk run as we understand it within this thesis. Milk runs are tours cyclically repeated according to a fixed schedule, with a fixed sequence and fixed arrival times, serving frequent supplier plant relations. The volumes are determined on a daily basis by a pull or push based order policy.

Regarding the regularity – and the resulting reliability – as a key characteristic of milk runs we pointed out the main advantages of milk runs and other transport concepts with regular components: better transport tariffs compared to ad hoc concepts, a reduced operational complexity, and a better planning ability for suppliers. Or stated more generally, the reduction of the variability by fixed schedules results in lower inventory, capacity, and time buffers within the supply network.

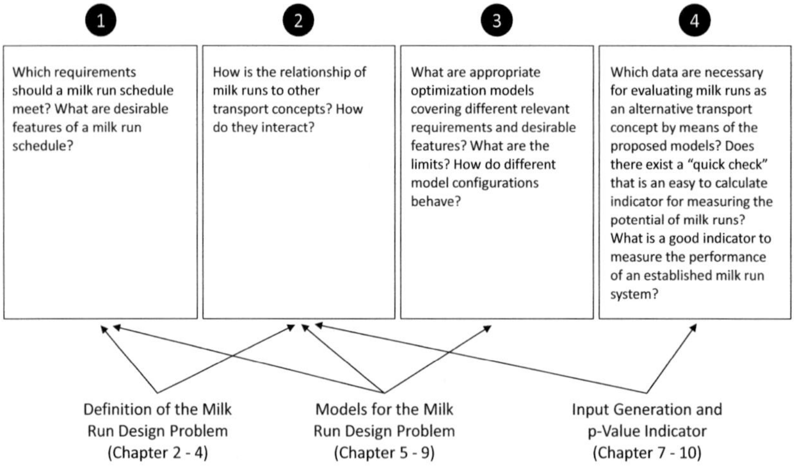

Figure 11.1: The key contributions of this thesis related to the four sets of research questions introduced in Chapter 1

Milk Runs and HEIJUNKA Levelled Supplier KANBAN Cycles

Besides reflecting milk runs as a general transport concept, we discussed the logic of supplier milk runs within lean manufacturing principles. In the lean world, milk run schedules are usually repeated on a daily basis. The parts are picked up at the respective supplier sites up to 8 times per day, and the cycle times for every supplier plant relation are approximately constant.

If milk runs are repeated on a weekly basis, the resulting inter arrival times for most frequencies are varying. Considering, for example, a weekly delivery profile Monday, Tuesday, Friday, the inter arrival times are one and two days respectively. If the demand is assumed to be constant, this results in varying lots. Since a levelled use of transport resources is a goal in lean manufacturing, we described in detail how HEIJUNKA levelling techniques can be applied to supplier KANBAN cycles in order to reach levelled transport lots in spite of inter arrival time varying delivery profiles. To the best of our knowledge there exists no description of such a system in lean manufacturing literature.

Milk Run Design

Ensuing from the previous results, we defined the milk run design problem as a tactical decision considering simultaneously the assignment of a transport concept and a frequency to every supplier plant relation, and the solution of the milk run scheduling problem for the suppliers assigned to the milk run concept. The objective is to minimize the cost for transports and inventory considering the capacity of vehicles and storage areas. Additional consistency features can be required to further reduce the operational complexity.

11.1.2 Models for the Milk Run Design Problem

On the basis of the definition of the milk run design problem we turned towards answering the set of research questions 3: We surveyed existing approaches from literature and proposed models addressing the problem in an integrated way. We evaluated different model configurations on small artificial instances and a real-world instance for addressing the research question 1 and 2 also from a modelling point of view.

Related Optimization Approaches

As a basis for our own models we gave an overview of related optimization approaches from literature coming from different research areas such as optimal lot sizing, network design, vehicle routing, and lean manufacturing. We classified the approaches by examining which transport concepts are considered and which types of regularity and consistency the resulting schedules show. Furthermore, we briefly introduced related VRP models.

From this overview the main research gaps with respect to the milk run design problem were derived: An explicit, deterministic model was missing which represents the inventory behaviour in case of cyclically repeated delivery profile schedules with levelled lot sizes. Furthermore, no tactical model in literature considered the transport concept assignment, the frequency assignment, and the milk run scheduling decision simultaneously. Consequently, no performance estimation concepts and performance indicator existed for the milk run concept relating it to the cost for alternative transport concepts.

Modelling the Milk Run Design Problem

For modelling the milk run design problem we proposed a decomposed inventory routing model considering consistency, outsourcing, and a heterogeneous fleet. We introduced different variants each addressing a specific aspect (see also Table 11.1).

The basic milk run scheduling model consists of an explicit deterministic cyclic inventory model and a set covering (partitioning) formulation of the periodic vehicle routing problem. In the PVRP, the results of the inventory model, namely a set of inventory optimal patterns for every frequency and the corresponding cost for every supplier, are taken as input. To restrict the deviation from the optimal inter arrival times assumed in the inventory model in the set cover part, the tours are explicitly scheduled. That means, in case of constant cycle schedules we add inter arrival time consistency constraints and in case of delivery profile schedules we add arrival time consistency constraints.

Another input for the set cover part are lower bounds for the frequency. They can be either determined by an economic lower bound or by the available storage capacities for every part. It can be left to the model to assign a higher frequency if that results in lower operational cost. As shown in Table 11.1, in case of two models we also considered driver and partial driver consistency.

Table 11.1: Characteristics covered by different variants of the proposed milk run design models

		MRS (Chapter 8)	TC-MRS (Chapter 9)	TC-MRS (Chapter 10)
Decision	nodes	–	✓	✓
	sequence	✓	✓	✓
	vehicle	✓	✓	✓
	period	✓	✓	✓
	frequency	✓	✓′	–
	volume	fixed by inventory model	fixed by inventory model	fixed by inventory model
	vehicle type	–	–	✓
Constraints	vehicle capacity	✓	✓	✓
	storage capacity	considered by inventory model	considered by inventory model	considered by inventory model
	arrival time cons.	✓	✓	–
	inter arrival time cons.	✓	–	–
	driver consistency	✓	✓	–
Objective	cost types	T (H) C	T (H) C	T

T	variable (and fixed) transport cost
H	inventory holding cost
O	outsourcing cost
C	penalty cost driver inconsistency

In the following we describe the model (components), the results of the computational experiments and the solution approach in more depth.

Cyclic Inventory Models for Pull and Push Systems As described above, the necessary input for the PVRP from the inventory model is a set of optimal delivery pattern for every frequency and the corresponding inventory cost. For deriving this input we extended deterministic finite horizon models from literature for both constant cycle schedules (continuous time inventory models) and delivery profile schedules (discrete time inventory models) to the cyclic variant considering the plant and the supplier site simultaneously.

Based on that, we worked out the different behaviour of pull and push based ordering systems in case of varying inter arrival times and varying transport lot sizes: Due to the varying lot sizes, the inventory level is asymmetrically distributed. In a pull case, the inventory at the plant site is always greater or equal to the inventory at the supplier site, while in push cases it is the other way around. From a supply chain management point of view the first variant, in which the parts are closer to their point of consumption, is more beneficial.

Since levelled lot sizes are favourable for reaching a levelled use of transport resources, we proposed a new model describing order policies in which the inter arrival times differ but the lot sizes are levelled. This can, for example, be reached by applying HEIJUNKA levelling techniques as described above. Since the optimality property for patterns cannot be expressed explicitly by common constraints for this policy, we proposed an algorithm determining the set of inventory optimal pattern efficiently.

We discussed the characteristics of the different models with respect to effects on the consistency of shipping volumes and the compatibility with time window restrictions of an inbound logistics network.

Milk Run Scheduling Considering Consistency Computational experiments on artificial instances showed that the decomposition of the inventory and routing part works well for delivery profile milk runs. On the test instances, both minimal inventory levels and low routing cost could be reached. The inventory cost are controlled by allowing only inventory optimal patterns and enforcing arrival time consistency, while the routing costs compared to the cost of the solution without asking for arrival time consistency are only slightly higher. This reconfirms the results reported by other authors for similar problem types on bigger instances.

To plan constant cycle milk runs, the proposed decomposition did not appear adequate. We have to enforce constant inter arrival times which leads to a significant routing cost increase for the test instances. This rise in transport cost is in most cases not compensated by considerably lower inventory cost. Hence, in case of constant cycle schedules an inventory routing model – explicitly considering the interaction of inventory and routing decisions – seems more appropriate. However, the experiments also showed that cyclic schedules are operationally complex in case of a cycle time of one week. The arrival times during the night and in the early morning are in conflict with night drive bans and with the opening hours of logistics providers and suppliers.

The experimental results furthermore showed that arrival time consistent delivery profile schedules are compatible with driver consistency. That means, the routing cost increase is in most cases moderate if all suppliers are always served by the same driver. In contrast, constant cycle schedules are in conflict with driver consistency. The cost increase in this case is significant.

Whenever driver consistency is not strictly necessary, the concept of partial driver consistency is very promising since a high level of consistency can be reached without increasing the routing cost. We could furthermore show that the same consistency level could be reached on our instances by applying a two stage approach: Instead of considering the partial driver consistency during the solution of the MRS, we reassign the tours in a second step in a way that the inconsistencies are minimized.

Transport Concept Assignment and Milk Run Scheduling Considering a Heterogeneous Fleet The extension of the MRS model to considering the assignment to a transport concept simultaneously was tested on both artificial instances and a real-world use case. As potential transport concepts we incorporated area forwarding, point-to-point transports, and milk runs – a typical setup in the automotive industry. On the artificial instances we investigated the behaviour in case of varying milk run cost parameters not being able to state something about cost savings. However, for the inbound network of a production site of Bosch automotive in Homburg the application of the TC-MRS model resulted in significant cost savings of up to 15% compared to the current situation in which all orders are sent by area forwarding services. This was especially surprising since the weight structure is – due to the low average weight of less than a ton – not obviously beneficial for a milk run system.

However, the high cost saving values between 12 and 15% are reached only for the instances considering 44 out of the 54 regular suppliers and all 54 suppliers, respectively, as milk run candidates. That means, the more visits are incorporated, the higher is the cost savings potential. We also showed on the use case data that there exist situations in which significant additional cost savings can be realised by considering vehicle classes with different cost and capacity parameters instead of a single vehicle type.

As a benchmark we also tested the performance of rule based transport concept assignment heuristics often used in practice: In this approach all orders below a certain weight threshold are sent through the AF network, the ones above are assigned to the milk run concept. The results showed that the heuristic approach yields significant cost savings but is very sensitive to the weight threshold. Furthermore, the TC-MRS resulted in all cases in higher or considerably higher cost savings.

A Priori Column Generation Approach A Priori Column Generation Approach: The goal of this thesis was to provide an adequate model for the milk run design problem and not necessarily a solution approach scaling to large instances. Therefore, we proposed a flexible a priori column generation approach, which could be applied to all model configurations and which yielded optimal results for nearly all models on all instances. We showed that the approach is limited since the number of feasible routes and the corresponding generation procedure grows exponentially with the number of suppliers. If the number of tours is high, also the solution process of the set cover or partitioning model is long. Both negative effects on the run time are reinforced for configurations in which time consistency is considered.

However, the TC-MRS model considering a heterogeneous fleet but no arrival time consistency could be solved to optimality for all instances of our real-world use case containing up to 54 suppliers within less than an hour. Other authors report on more producing companies having only between 20 and up to 100 relevant, regular suppliers. Hence, the approach is applicable at least to smaller real-world instances yielding optimal results.

11.1.3 Input Generation and p-Value Indicator

The set of research questions 4 is treated within the case study. However, the results are generally applicable.

The procedures to generate the necessary input data for solving a milk run design model are transferable to other companies since they are based on historical data relevant to billing. These data are easily accessible also for small and mid sized companies.

Solving the milk run design problem based on the derived input parameters offers an upper bound for potential savings. This upper bound can be used to decide if more detailed, usually difficult to gather, data such as net demands and packaging details should be collected. These data are needed to check if the order policy can be adapted in a way that the frequency and the shipping volumes are sufficiently stable.

We furthermore introduced the – generally applicable – p-value indicator defined by the share of AF cost for an order and the milk run cost accrued if the corresponding supplier is served on an exclusive milk run. These cost values can be easily determined without solving a vehicle routing problem. We showed that the maximum p-value over all suppliers can be used to approximately assess the potential of milk runs: If it is considerably below 0.8, it is not very likely that milk runs are a cost efficient alternative to the area forwarding network. If the value is above 0.8 – according to our results – it seems to be worth to evaluate the savings potential. Whenever the value is above one, at least one milk run results in cost savings and an in depth evaluation of the concept seems promising. More generally, the p-value can be also used to assess AF tariff tables compared to exclusive milk runs or dedicated ad hoc tours.

The p-value on tour cluster level describes the share of virtual AF cost to the actual milk run cost including potential extra cost accrued if the demand exceeds the capacity of the vehicle. It is a good cost efficiency indicator measuring the performance of a milk run cluster with respect to alternative transport concepts. If it is continuously below one, the milk run should be adapted.

The experiments in which we varied the milk run cost showed that the better the AF tariffs the fewer milk runs are cost efficient but the better the milk runs are in terms of capacity usage. This shows that solving the milk run design problem corresponds to "cherry picking" with respect to the freight forwarders. If in the long run, milk runs are outsourced to differing freight forwarders than the area forwarding service provider, this might result in higher AF tariffs. This effect should be considered when contracts with logistics providers are arranged.

11.2 Outlook

The results of our experiments showed that the TC-MRS is adequate to model the milk run design decision if the receiving plant aims for weekly recurring transport schedules. However, different extensions enlarge the scope of application scenarios: If the TC-MRS is extended by considering multi echelons, milk runs consisting of a pre- and a main-leg or a pre-, main- and sub-leg can be built. This is especially interesting if a group of neighbouring frequent suppliers is located far away from the receiving plant. In case of significant reverse flows of parts to suppliers – for example in case of metal finishing processes executed by a supplier – the routing model should consider pick ups and deliveries simultaneously.

If the AF tariffs differentiate between pre- and main-leg tables, the AF cost cannot be determined before the orders are assigned to days since the tariffs for the main-leg depend on the resulting volumes per day. In this case the TC-MRS model must simultaneously cover this decision.

Even if it is assumed that sufficiently stable frequencies and volumes can be reached in general, in practice the capacity of milk runs will be exceeded on days with a higher demand and the capacity usage rate will be low on days with lower demand. This effect is reinforced by the fact that the demands of different parts for the same plant are often correlated. For cases of a stronger variability, this stochastic behaviour should be incorporated in the model considering different possible recourse strategies for exceeding demands.

For solving bigger instances, other exact or heuristic approaches need to be developed. From the literature overview and the results of this thesis we want to point out different possible directions: A column generation approach with exact or heuristic subprocedures seems promising. However, in both cases the time synchronisation resulting from time consistency requirements complicates the approach since one must assure that compatible tours with respect to the dynamic time windows are generated.

A heuristic approach successfully applied to other vehicle routing problems with synchronisation and to a broad range of rich, real-world vehicle routing problems is the adaptive large neighbourhood search. Such an approach will most probably generate good results in short time. The p-value might be a helpful indicator for selecting suppliers for relaxing the solution and for sorting them for the reinsertion.

References

Akturk, M. and F. Erhun (1999). An overview of design and operational issues of kanban systems. *International Journal of Production Research 37*(17), 3859–3881.

Alegre, J., M. Laguna, and J. Pacheco (2007). Optimizing the periodic pick-up of raw materials for a manufacturer of auto parts. *European Journal of Operational Research 179*, 736–746.

Andersson, H., J. M. Duesund, and K. Fagerholt (2011). Ship routing and scheduling with cargo coupling and synchronization constraints. *Computers and Industrial Engineering 61*, 1107–1116.

Andersson, H., A. Hoff, M. Christiansen, G. Hasle, and A. Løkketangen (2010). Industrial aspects and literature survey: Combined inventory management and routing. *Computers & Operations Research 37*(9), 1515–1536.

Archetti, C. and M. Speranza (2008). The split delivery vehicle routing problem: A survey. In B. Golden, S. Raghavan, and E. Wasil (Eds.), *The Vehicle Routing Problem: Latest Advances and New Challenges*, Volume 43 of *Operations Research/Computer Science Interfaces*, pp. 103–122. Springer.

Bahinipati, B. K., A. Kanda, and S. Deshmukh (2009). Coordinated supply management: review, insights, and limitations. *International Journal of Logistics: Research and Applications 12*(6), 407–422.

Baldacci, R., M. Battarra, and D. Vigo (2008). Routing a heterogeneous fleet of vehicles. In B. Golden, S. Raghavan, and E. Wasil (Eds.), *The Vehicle Routing Problem: Latest Advances and New Challenges*, Volume 43 of *Operations Research/Computer Science Interfaces*, pp. 3–27. Springer.

Baudin, M. (2004). *Lean Logistics: the nuts and bolts of delivering materials and goods.* Productivity Press.

Berman, O. and Q. Wang (2006). Inbound logistic planning: minimizing transportation and inventory cost. *Transportation science 40*(3), 287–299.

Bicheno, J. and M. Holweg (2009). *The lean toolbox: The essential guide to lean transformation.* Production and Inventory Control, Systems and Industrial Engineering (PICSIE) Books.

Birbil, Ş. İ., K. Bülbül, H. Frenk, and H. Mulder (2014). On eoq cost models with arbitrary purchase and transportation costs. *Journal of Industrial and Management Optimization.*

Blumenfeld, D. E., L. D. Burns, C. F. Daganzo, M. C. Frick, and R. W. Hall (1987). Reducing logistics costs at general motors. *Interfaces 17*(1), 26–47.

Blumenfeld, D. E., L. D. Burns, J. D. Diltz, and C. F. Daganzo (1985). Analyzing trade-offs between transportation, inventory and production costs on freight networks. *Transportation Research Part B: Methodological 19*(5), 361–380.

Böhle, C. and W. Dangelmaier (2009). Milk run optimization with delivery windows and hedging against uncertainty. In *Operations Research Proceedings 2008*, pp. 247–252. Springer.

Boysen, N., S. Emde, M. Hoeck, and M. Kauderer (2015). Part logistics in the automotive industry: Decision problems, literature review and research agenda. *European Journal of Operational Research 242*(1), 107–120.

Branke, J., D. Häußler, and C. Schmidt (2007). Transport channel selection. In *Operations Research Proceedings 2006*, pp. 349–354. Springer.

Breier, H. and T. Gossler (2014). Branch-and-price on the split delivery vehicle routing problem with time windows and alternative delivery periods. In *Operations Research Proceedings 2013*, pp. 57–65. Springer.

Burns, L. D., R. W. Hall, D. E. Blumenfeld, and C. F. Daganzo (1985). Distribution strategies that minimize transportation and inventory costs. *Operations Research 33*(3), 469–490.

Campbell, A., L. Clarke, A. Kleywegt, and M. Savelsbergh (1998). The inventory routing problem. In T. Crainic and G. Laporte (Eds.), *Fleet Management and Logistics.* Springer.

Chan, H. K. and F. T. Chan (2010). A review of coordination studies in the context of supply chain dynamics. *International Journal of Production Research 48*(10), 2793–2819.

Chen, Z. and B. R. Sarker (2014). An integrated optimal inventory lot-sizing and vehicle-routing model for a multisupplier single-assembler system with jit delivery. *International Journal of Production Research 52*(17), 5086–5114.

Chopra, S. and P. Meindl (2013). *Supply chain management: strategy, planning, and operation* (5. ed., global ed. ed.)., Chapter Transportation in a Supply Chain, pp. 409–439. Pearson.

Chuah, K. and J. Yingling (2005). Routing for a just-in-time supply pickup and delivery system. *Transportation Science 39*(3), 328–339.

Coelho, L. C., J.-F. Cordeau, and G. Laporte (2012). Consistency in multi-vehicle inventory-routing. *Transportation Research Part C: Emerging Technologies 24*, 270–287.

Dekker, R., J. Bloemhof, and I. Mallidis (2012). Operations research for green logistics–an overview of aspects, issues, contributions and challenges. *European Journal of Operational Research 219*(3), 671–679.

Dickmann, P. (2009). *Schlanker Materialfluss: mit Lean Production, Kanban und Innovationen*, Volume 2nd edition, Chapter Die KANBAN Steuerung, pp. 162–168. Springer.

Drexl, M. (2012). Synchronization in vehicle routing – a survey of vrps with multiple synchronization constraints. *Transportation Science 46*, 297–316.

Dyer, J. H. (1996). Specialized supplier networks as a source of competitive advantage: Evidence from the auto industry. *Strategic management journal 17*(4), 271–291.

Ekici, A., O. Ö. Özener, and G. Kuyzu (2014). Cyclic delivery schedules for an inventory routing problem. *Transportation Science Articles in Advance.*

Esparcia-Alcázar, A. I., M. Cardós, J. Merelo, A. Martínez-García, P. García-Sánchez, E. Alfaro-Cid, and K. Sharman (2009). Evita: An integral evolutionary methodology for the inventory and transportation problem. In *Bio-inspired Algorithms for the Vehicle Routing Problem*, pp. 151–172. Springer.

European Parliament and Council (2006, 04). Regulation (ec) no 561/2006 of the european parliament and of the council of 15 march 2006 on the harmonisation of certain social legislation relating to road transport and amending council regulations (eec) no 3821/85 and (ec) no 2135/98 and repealing council regulation (eec) no 3820/85.

Feillet, D., T. Garaix, F. Lehuédé, O. Péton, and D. Quadri (2010). A new consistent vehicle routing problem for the transportation of handicapped persons. *Working Paper EMSE CMP-SFL 9.*

Feillet, D., T. Garaix, F. Lehuédé, O. Péton, and D. Quadri (2014). A new consistent vehicle routing problem for the transportation of people with disabilities. *Networks 63*(3), 211–224.

Fleischmann, B. (1999). Transport and inventory planning with discrete shipment times. In *New Trends in Distribution Logistics*, pp. 159–178. Springer.

Francis, P., K. Smilowitz, and M. Tzur (2006a). Flexibility and complexity in periodic distribution problems. *Naval Research Logistics 54*(2), 136–150.

Francis, P., K. Smilowitz, and M. Tzur (2006b). The period vehicle routing problem with service choice. *Transportation Science 40*(4), 439–454.

Francis, P., K. Smilowitz, and M. Tzur (2008). The period vehicle routing problem and its extensions. In B. Golden, S. Raghavan, and E. Wasil (Eds.), *The Vehicle Routing Problem: Latest Advances and New Challenges*, Volume 43 of *Operations Research/Computer Science Interfaces*, pp. 73–102. Springer.

Furmans, K. and M. Veit (2013). Models of leveling for lean manufacturing systems. In *Handbook of Stochastic Models and Analysis of Manufacturing System Operations*, pp. 115–138. Springer.

Gaur, V. and M. Fisher (2004). A periodic inventory routing problem at a supermarket chain. *Operations Research 52*, 813–822.

Glock, C. H. (2012). The joint economic lot size problem: A review. *International Journal of Production Economics 135*(2), 671–686.

Groër, C., B. Golden, and E. Wasil (2009). The consistent vehicle routing problem. *Manufacturing & service operations management 11*(4), 630–643.

Grunewald, M., T. Volling, and T. S. Spengler (2011). Towards leveled inventory routing for the automotive industry. In *Operations Research Proceedings 2010*, pp. 423–428. Springer.

Gudehus, T. (2012). Logistikkosten und Leistungskosten. In *Logistik 1, Grundlagen, Verfahren und Strategien.* Springer.

Harks, T., F. G. König, J. Matuschke, A. T. Richter, and J. Schulz (2014). An integrated approach to tactical transportation planning in logistics networks. *Transportation Science Articles in Advance.*

Harris, F. W. (1990). How many parts to make at once. *Operations Research 38*(6), 947–950.

Heckmann, I. (2015). *Towards Supply Chain Risk Analytics – Fundamentals, Simulation, Optimization.* Ph. D. thesis, Karlsruher Institut für Technologie. To be published.

Hines, P., M. Holweg, and N. Rich (2004). Learning to evolve: a review of contemporary lean thinking. *International Journal of Operations & Production Management 24*(10), 994–1011.

Hopp, W. and M. Spearman (2001a). *Factory Physics. Foundations of Manufacturing Management*, Volume Second Edition. McGraw-Hill Irwin.

Hopp, W. and M. Spearman (2001b). *Factory Physics. Foundations of Manufacturing Management*, Volume Second Edition, Chapter Inventory Control: From EOQ to ROP, pp. 48–108. McGraw-Hill Irwin.

Hopp, W. J. (2011). *Supply chain science.* Waveland Press.

Hosseini, S. D., M. A. Shirazi, and S. M. T. F. Ghomi (2014). Harmony search optimization algorithm for a novel transportation problem in a consolidation network. *Engineering Optimization 46*(11), 1538–1552.

Hosseini, S. D., M. A. Shirazi, and B. Karimi (2014). Cross-docking and milk run logistics in a consolidation network: A hybrid of harmony search and simulated annealing approach. *Journal of Manufacturing Systems 33*(4), 567–577.

Hsiao, H., R. G. Kemp, J. Van der Vorst, and S. Omta (2010). A classification of logistic outsourcing levels and their impact on service performance: Evidence from the food processing industry. *International Journal of Production Economics 124*(1), 75–86.

Jafari-Eskandari, M., A. Aliahmadi, and G. Khaleghi (2010). A robust optimisation approach for the milk run problem with time windows with inventory uncertainty: an auto industry supply chain case study. *International Journal of Rapid Manufacturing 1*(3), 334–347.

Jaruphongsa, W., S. Çetinkaya, and C.-Y. Lee (2007). Outbound shipment mode considerations for integrated inventory and delivery lot-sizing decisions. *Operations Research Letters 35*(6), 813–822.

Kempkes, J. P. (2009). *Kostenoptimale Materialfluüsse in der operativen Zulieferungslogistik der Nutzfahrzeugindustrie.* Ph. D. thesis, Universität Paderborn.

Kempkes, J. P., A. Koberstein, and L. Suhl (2010). A resource based mixed integer modelling approach for integrated operational logistics planning. In *Advanced Manufacturing and Sustainable Logistics*, pp. 281–294. Springer.

Khouja, M. and S. Goyal (2008). A review of the joint replenishment problem literature: 1989–2005. *European Journal of Operational Research 186*(1), 1–16.

Kilic, H. S., M. B. Durmusoglu, and M. Baskak (2012). Classification and modeling for in-plant milk-run distribution systems. *The International Journal of Advanced Manufacturing Technology 62*, 1135–1146.

Klug, F. (2010a). *Logistikmanagement in der Automobilindustrie: Grundlagen der Logistik im Automobilbau*, Chapter Aufgabenbereiche der Logistikplanung, pp. 149–252. Springer.

Klug, F. (2010b). *Logistikmanagement in der Automobilindustrie: Grundlagen der Logistik im Automobilbau*, Chapter Lean Logistics, pp. 253–285. Springer.

Konur, D. and A. Toptal (2012). Analysis and applications of replenishment problems under stepwise transportation costs and generalized wholesale prices. *International Journal of Production Economics 140*(1), 521–529.

Kotani, S. (2007). Optimal method for changing the number of kanbans in the e-kanban system and its applications. *International Journal of Production Research 45*(24), 5789–5809.

Kovacs, A. A., B. L. Golden, R. F. Hartl, and S. N. Parragh (2014). The generalized consistent vehicle routing problem. *Transportation Science Articles in Advance.*

Kovacs, A. A., S. N. Parragh, and R. F. Hartl (2014). A template-based adaptive large neighborhood search for the consistent vehicle routing problem. *Networks 63*(1), 60–81.

Kovalev, A. and C. Ng (2008). A discrete eoq problem is solvable in o (logn) time. *European Journal of Operational Research 189*(3), 914–919.

Krajewska, M. A. (2008). *Potentials for efficiency increase in modern freight forwarding.* Springer.

Kuhn, H. and T. Liske (2014). An exact algorithm for solving the economic lot and supply scheduling problem using a power-of-two policy. *Computers & Operations Research 51*, 30–40.

Kumar, C. S. and R. Panneerselvam (2007). Literature review of jit-kanban system. *The International Journal of Advanced Manufacturing Technology 32*(3-4), 393–408.

Li, F., B. Golden, and E. Wasil (2007). The open vehicle routing problem: Algorithms, large-scale test problems, and computational results. *Computers & Operations Research 34*(10), 2918–2930.

Liker, J. K. (2004). *The toyota way.* Esensi.

Lippolt, C. R. and K. Furmans (2008). Sizing of heijunka-controlled production systems with unreliable production processes. In *Lean business systems and beyond*, pp. 11–19. Springer.

Lourenço, H. R. and R. Ribeiro (2011). Strategies for an integrated distribution problem. *Hybrid algorithms for service, computing and manufacturing systems: routing, scheduling and availability solutions.*, 98–121.

Matzka, J., M. Di Mascolo, and K. Furmans (2012). Buffer sizing of a heijunka kanban system. *Journal of Intelligent Manufacturing 23*(1), 49–60.

Meixel, M. and M. Norbis (2008). A review of the transportation mode choice and carrier selection literature. *International Journal of Logistics Management 19*(2), 183–211.

Meyer, A., A. Cardeneo, and K. Furmans (2011). A two stage approach for balancing a periodic long-haul transportation network. In *Operations Research Proceedings 2010*, pp. 257–262. Springer.

Miemczyk, J. and M. Holweg (2004). Building cars to customer order - what does it mean for inbound logistics operations? *Journal of Business Logistics 25*(2), 171–197.

Miyazaki, S., H. Ohta, and N. Nishiyama (1988). The optimal operation planning of kanban to minimize the total operation cost. *The International Journal Of Production Research 26*(10), 1605–1611.

Monden, Y. (2012). *Toyota production system: an integrated approach to just-in-time* (4. ed.). Boca Raton: CRC Press.

Mourgaya, M. and F. Vanderbeck (2007). Column generation based heuristic for tactical planning in multi-period vehicle routing. *European Journal of Operational Research 183*(3), 1028–1041.

Natarajarathinam, M., J. Stacey, and C. Sox (2012). Near-optimal heuristics and managerial insights for the storage constrained, inbound inventory routing problem. *International Journal of Physical Distribution & Logistics Management 42*(2), 152–173.

Ohlmann, J., M. Frey, and T. Barrett (2008). Route design for lean production systems. *Transportation Science 42*(3), 352–370.

Ohno, K., K. Nakashima, and M. Kojima (1995). Optimal numbers of two kinds of kanbans in a jit production system. *The International Journal of Production Research 33*(5), 1387–1401.

Parthanadee, P. and R. Logendran (2006). Periodic product distribution from multi-depots under limited supplies. *IIE Transactions 38*(11), 1009–1026.

Peterson, B., W.-J. van Hoeve, S. Kekre, and L. Debo (2010). Flexible milk-runs for stochastic vehicle routing. *Working Paper Tepper School Of Business.*

Piplani, R. and S. Viswanathan (2004). Supply chain inventory coordination through multiple, common replenishment epochs and selective discount. *International Journal of Logistics Research and Applications 7*(2), 109–118.

Projektgruppe Standardbelieferungsformen (2008). Standardbelieferungsformen der Logisitik in der Automobilindustrie. In *Recommondation 5051.* VDA - German Association of the Automotive Industry.

Queiser, H. (2007). Anlieferkonzepte in der Automobilindustrie - Ein internationaler Vergleich. In *Dokumentation – Zukunft AutomobilMontage.* ATZ/MTZ–Konferenz Automobil-Montage 2007. `http://www.4flow.de/fileadmin/user_upload_publikationen/veroeffentlichungen/Anlieferkonzepte_in_der_Automobilindustrie_Dr._Hartmut_Queiser.pdf` (17.02.2015).

Quélin, B. and F. Duhamel (2003). Bringing together strategic outsourcing and corporate strategy: Outsourcing motives and risks. *European management journal 21*(5), 647–661.

Raa, B. and E.-H. Aghezzaf (2008). Designing distribution patterns for long-term inventory routing with constant demand rates. *International Journal of Production Economics 112*(1), 255–263.

Raa, B. and E.-H. Aghezzaf (2009). A practical solution approach for the cyclic inventory routing problem. *European Journal of Operational Research 192*(2), 429–441.

Ratliff, H. D. and W. G. Nulty (1997). *Logistics composite modeling.* Springer.

Rieksts, B. Q. and J. A. Ventura (2008). Optimal inventory policies with two modes of freight transportation. *European Journal of Operational Research 186*(2), 576–585.

Robinson, P., A. Narayanan, and F. Sahin (2009). Coordinated deterministic dynamic demand lot-sizing problem: A review of models and algorithms. *Omega 37*(1), 3–15.

Rother, M. and J. Shook (1999). *Learning to see: value stream mapping to add value and eliminate muda.* Productivity Press.

Rusdiansyah, A. and D.-b. Tsao (2005). An integrated model of the periodic delivery problems for vending-machine supply chains. *Journal of Food Engineering 70*(3), 421–434.

Sadjadi, S., M. Jafari, and T. Amini (2009). A new mathematical modeling and a genetic algorithm search for milk run problem (an auto industry supply chain case study). *The International Journal of Advanced Manufacturing Technology 44*(1-2), 194–200.

Schmid, V., K. F. Doerner, and G. Laporte (2013). Rich routing problems arising in supply chain management. *European Journal of Operational Research 224*(3), 435–448.

Schöneberg, T., A. Koberstein, and L. Suhl (2010). An optimization model for automated selection of economic and ecologic delivery profiles in area forwarding based inbound logistics networks. *Flexible services and manufacturing journal 22*(3-4), 214–235.

Schöneberg, T., A. Koberstein, and L. Suhl (2013). A stochastic programming approach to determine robust delivery profiles in area forwarding inbound logistics networks. *OR spectrum 35*(4), 807–834.

Simchi-Levi, D., X. Chen, and J. Bramel (2014a). Economic lot size models with constant demands. In *The Logic of Logistics*, Springer Series in Operations Research and Financial Engineering, pp. 137–150. Springer New York.

Simchi-Levi, D., X. Chen, and J. Bramel (2014b). Economic lot size models with varying demands. In *The Logic of Logistics*, Springer Series in Operations Research and Financial Engineering, pp. 137–150. Springer New York.

Smilowitz, K., M. Nowak, and T. Jiang (2013). Workforce management in periodic delivery operations. *Transportation Science 47*(2), 214–230.

Speranza, M. G. and W. Ukovich (1994). Minimizing transportation and inventory costs for several products on a single link. *Operations Research 42*(5), 879–894.

Spliet, R. (2013). *Vehicle Routing with Uncertain Demand.* Ph. D. thesis, Erasmus University Rotterdam.

Stacey, J., M. Natarajarathinam, and C. Sox (2007). The storage constrained, inbound inventory routing problem. *International Journal of Physical Distribution & Logistics Management 37*(6), 484–500.

Stenger, A., M. Schneider, and D. Goeke (2013). The prize-collecting vehicle routing problem with single and multiple depots and non-linear cost. *EURO Journal on Transportation and Logistics 2*(1-2), 57–87.

Swenseth, S. R. and M. R. Godfrey (2002). Incorporating transportation costs into inventory replenishment decisions. *International Journal of Production Economics 77*(2), 113–130.

Tempelmeier, H. (2007). On the stochastic uncapacitated dynamic single-item lotsizing problem with service level constraints. *European Journal of Operational Research 181*(1), 184–194.

Viswanathan, S. and R. Piplani (2001). Coordinating supply chain inventories through common replenishment epochs. *European Journal of Operational Research 129*(2), 277–286.

Wang, X. (2015). From cherry-picking to integrated operational transportation planning. In *Operational Transportation Planning of Modern Freight Forwarding Companies*, pp. 43–56. Springer.

Womack, J. P., D. T. Jones, and D. Roos (1990). *The Machine That Changed the World: The Story of Lean Production.* Rawson Associates.